Agricultural Education Status & Prospects

Agricultural Education Status & Prospects

Authors

Dr. Lokesh G.B.

Assistant Professor, Department of Agril. Economics,
College of Agriculture, UAS, Raichur, Karnataka

Dr. D.M. Chandargi

Professor of Agril., Extension Education & University Librarian,
UAS, Raichur, Karnataka

Dr. M.G. Chandrkanth

Professor of Agril., Economics & Director
Institute for Social and Economics Change (ISEC) Bangalore, Karnataka

Dr. M.V. Ravi

Assistant Professor, Department of Soil Science & Agril., Chemistry,
College of Agriculture, UAS, Raichur, Karnataka

International Books & Periodical Supply Service

(Publisher of Scientific Books)

38, NISHANT KUNJ, PITAM PURA MAIN ROAD
DELHI-110034 (India)
Phone : 011-27352078, Mobile : 09810146811, E-mail : hkjain1975@yahoo.com

Published by :

International Books & Periodical Supply Service

(Publisher of Scientific Books)
38, NISHANT KUNJ, PITAM PURA MAIN ROAD
DELHI-110034 (India)
Phone : 011-27352078, Mobile : 09810146811, E-mail : hkjain1975@yahoo.com

ISBN 978-93-88892-03-2

Composed, Designed & Printed in India

About the Authors

Dr. Lokesh G.B., is the Assistant Professor, Department of Agricultural Economics, UAS, Raichur. He is a recipient of two International Fellowship awards for attending international courses: 1. Exploration, Exploitation and Management of Groundwater Resources at Hebrew University of Jerusalem, Israel sponsored by Mashav, Foreign Ministry, Govt. of Israel (2003) 2. NAIP-ICAR International Fellowship for attending international training on Commodity Markets at IFPRI, Washington DC, USA (2013) awarded by NAIP-ICAR, New Delhi, Govt. of India. He has successfully completed two national and four Govt. of Karnataka funded research projects on various issues related Agricultural Policy. He also teaches courses in Agricultural Economics for both undergraduate post graduate students at College of Agriculture, Raichur since 2011. Prior to his teaching and research experiences in agriculture, he has also worked 5 years in corporate sector in providing analytical solutions for solving the business problems in fast moving consumer's goods, software marketing and retail financing sectors at global level, etc. To his credit totally 15 research articles published in reputed national and international journals and three in book, in chapters. He also contributed in the form presentation and participation in national and international conferences. E-mail: gblokesh@yahoo.com

Dr. D.M. Chandargi, completed Ph.D.in Agricultural Extension Education in the year 1996 from University of Agricultural Sciences, Dharwad, Karnataka. He has put in around 35 years of service in different capacities as extension worker,scientist and administrator in State Department of Agriculture and Agricultural University. Now, he is serving as Dean (Agri.) at College of Agriculture, Raichur under the University of Agricultural Sciences, Raichur, Karnataka. He has also served as a Registrar of University of Agricultural Sciences, Raichur for a period of two and half years. He has significantly contributed to the field of Agricultural extension education, management, rural development and sustainable livelihoods in particular. He has worked on income generation activities of SHGs and evolved policy document through world bank funded Sujala watershed project. He has worked as outreach expert in international project on "Consolidation of Food Security in South India" funded by CIDA collaborated by Mc Gill University, Canada.

Dr. D.M. Chandargi, is a recipient of many awards, some of them are:

- University of Agricultural Sciences, Gold Medal and Dr.R. Dwarakinath Gold medal for Ph.D. Degree programme.
- Best Paper Presentation Award - 2013 by International Society of Extension Education.
- Scientist of the Year Award - 2016 by National Environmental Science Academy, New Delhi.
- Excellence in Communication Award - 2016 by Society for Scientific Development in Agriculture and Technology, India.
- Outstanding Scientist in Agriculture Award - 2017 by Scientific Education Research Society, Meerut (U.P.).

As a teacher he has taught several courses of undergraduate and post graduate programmes including Ph.D. and guided twenty five M.Sc. and three Ph.D. students as a Chairman and more than 75 students as a member of advisory committees.

M.G. Chandrakanth is currently Director of Institute for Social and Economic Change (ISEC), Nagarabhavi, Bangalore. He was Professor and University Head of Agricultural Economics and former Dean of College of Agriculture, University of Agricultural Sciences, Bangalore (UASB), India. He completed B.Sc. (Agri.) in 1974, M.Sc. (Agri.) in Agricultural Economics in 1976 under Professor Noel SP Rebellow, and obtained his Ph.D. in Agricultural Economics in 1981 under Dr. JV Venkataram from the UASB. He served the Institute for Social and Economic Change, Bangalore, India, as an Assistant Professor in the Rural Economics Unit from 1982-1983 and joined the UASB as an Associate Professor in 1983. He has been teaching agricultural economics, production economics, mathematical economics, advanced micro-economics, institutional economics and history of economic thought, natural resource and environmental economics courses.

Dr. Chandrakanth pursued post doctoral studies with fellowship from the Ford Foundation, USA in forest resource economics in 1987-88. He was awarded the Ciriacy-Wantrup Post Doctoral Fellowship for 1988-90 period, in institutional economics of natural resources at the University of California, Berkeley, USA and worked under Prof. Jeff Romm specializing in water resource economics and policy.

He was nominated as Vice President of the Indian Society of Agricultural Economics in 2000 and as a member of Editorial board of the *Indian Journal of Agricultural* Economics in 2012. He facilitated student exchange program with the University of Gent with Prof. Guido Van Huylenbroeck in Rural Development (IMRD) and was Erasmus Mundus visiting scholar, University of Ghent, Belgium in 2008; DAAD scholar at Humboldt University, Berlin 2010 and 2012. He is a member of the Water Policy Law Group of the University of South Australia, Adelaide, Australia.

Dr. Chandrakanth was awarded the Fellow of the Indian Society of Agricultural Economics in 2013 and *Dr. Ramesh Chandra Agrawal Award of Excellence* for his outstanding contributions in Agricultural Economics in 2014. He has published in *Natural Resources Journal, Natural Resources Forum, Ecological Economics, Water Policy, Environment and Development Economics, Economic and Political Weekly, Ind Jr of Agri Economics, Agricultural Economics Research Review* (http://www.toenre.com). He handled projects funded by the Ford Foundation from 1995 to 2005 on Groundwater resource economics and has been the leader of the Team of Excellence in Natural Resource Economics since 2000.

Dr. M.V. Ravi working as an Assistant Professor in the Department of Soil Science & Agricultural Chemistry, University of Agricultural Sciences, Raichur, Karnataka, India. He has completed B.Sc. (Agri.), M.Sc. (Agri.) and Ph.D. at University of Agricultural Sciences, Bangaluru, Karnataka, India. He had an experience of about 15 years in the field of Soil Science & Agricultural Chemistry. From past ten years, he is working as an Assistant Professor and teaching various Undergraduate and Postgraduate courses in the University. He has published nearly 75 scientific articles in various scientific journals at national as well as university level and presented various papers in international/national and local seminars, symposia, conferences, workshop, etc. He served as Staff Advisor, Liaison Officer, Nodal Officer, Technical Assistant to the Dean (PGs), Registrar and DOE in the university in technical administration for upliftment of the university/institution. He has conducted more than 70 training programme in Bhoochetana programme. which is a flagship project of Govt. of Karnataka and delivered more than 100 Guest lecturer.

Preface

Education is the process of receiving or giving systematic instruction, especially at a school or university. Agricultural Education is the teaching of agriculture, natural resources, and land management through hands on experience and guidance to prepare students for entry level jobs or to further education to prepare them for advanced agricultural jobs. Human Resource Development (HRD) is the framework for helping students/ employees to develop their personal and organizational skills, knowledge and abilities. Trained human resources are needed in different sectors of agriculture for targeted growth. The sectors are to be serviced by human resources with higher skills than before to ensure technology generation, its transfer to and more importantly its application at the grassroot level. Further, agricultural human resources with diverse skills is today required by a wide- ranging and fast expanding food processing industry, corporate and unorganized sector, etc.

Approach. To meet the objectives of the study, the available quantitative data was collected from all farm universities, annual reports of development departments. Stakeholder survey was also carried out in order assess the quality of agriculture education in the state. Compound Annual Growth Rate (CGAR) and System Dynamic model approach was used for forecasting supply and demand for agriculture and allied sectors graduates.

Features. This book contains a number of features to enhance knowledge, information and skills requirements in agricultural sector for better decision-making by policy-makers, entrepreneurs and students. The forecasting of future agricultural human resources requirements assumes considerable significance in identifying the trends and visualizing the emerging technologies and the corresponding needs for the appropriate human resources mix. The assessment of existing human resources, the sectoral growth and capacity utilisation of qualified graduates is indispensable to envisage the future agricultural education needs.

Organization. To provide a foundation, Chapters 1–2 describe agriculture scenario and its performance in Karnataka, which gives basic understanding of agricultural education, growth and development of agricultural sector in Karnataka. Chapter 3 explains the approach and methodology adopted in assessing and estimation of human resources requirements in agricultural sector. Chapters 4 to 9 give you about human resources assessment in agricultural and allied sectors. Chapter 10 is about overall demand & Supply and gap analysis chapter 11 and 12 describes about review of schemes and programmes and their quality assessment for making students friendly learning process, and Chapter 13 explains the issues, challenges and recommendations for improving the agricultural education to meet the food production demand in future.

Authors:
Dr. Lokesh G.B.
Dr. D.M. Chandargi
Dr. M.G. Chandrkanth
Dr. M.V. Ravi

Acknowledgements

This book is from the project "Assessment and Estimation of Human Resource Requirements in Agriculture and Allied Sectors in Karnataka" funded by HRD division, ICAR, New Delhi. We would like to thank Deputy Director General (Education), and Assistant Director General (ADG), HRD, ICAR who gave us this important project and helped us from time to time in executing the same. In particular, we are thankful to Dr. M. B. Chatti, ADG (HRD), Education division, ICAR. We are grateful to Dr. Narendra Singh Rathore, DDG (Edn.), ICAR for giving opportunity to conduct this study.

We thank all of them, mainly the contributions made by a number of organizations/companies/institutions, individual entrepreneurs, education experts, students, teachers and so on. Data has been collected from all farm universities of the Karnataka state, colleges, industrial establishments, government departments, students, teachers, NGOs, and other experts-we are grateful to all of them. A number of organizations/ institutions helped us in organization of Focus Group Discussions across the state, without whose support it would have not been possible for us to complete the work.

We gratefully acknowledge the suggestions made by Dr. D.M. Chandargi, University Librarian, UAS, Raichur and Dr. M G Chandrakanth, Director, Institute for Socio Economic Change (ISEC), Nagarabhavi, Bangalore helped in different phases of the execution of the work.

I sincerely thank Dr. K.N. Kattimani, Honourable Vice Chancellor, UAS, Raichur, and Dr. S.K. Meti, Director of Education, Dr. I. Shankergoud, Dean (PGS) and Dr. M.G. Patil, Dean (Agri.), College of Agriculture, UAS, Raichur for their encouragement, support and suggestion in improving the quality of the book.

Contents

Chapter 1

Introduction

1.1 Agriculture Scenario in Karnataka

Karnataka state is blessed with diversified agro-climatic conditions which are suitable for cultivation of various crops. The contribution of agricultural sector to GSDP is accounted for 12 per cent during 2016-2017. About 52.13 per cent of Karnataka state population relies on agriculture as their primary means of livelihood. Agriculture is characterised by wide crop diversification and remains dependent on vagaries of monsoon. Karnataka agriculture is constrained by large rainfed areas, low irrigation facilities, low value, low yield dominant cropping pattern and a large share of dependent population as evidenced by nearly two-third of the Gross Cropped Area (GCA) in the state in 12 districts of dry northern region, where only 30% of the area is irrigated.

Landuse pattern in the State (2014-15) indicates that the net sown area (NSA) constituted 100.44 lakh ha, followed by forest 30.73 lakh ha, the land not available for cultivation, including land put to non-agriculture uses, barren 7.87 lakh ha & uncultivable land 14.61 lakh ha, other uncultivated land including cultivable waste 4.09 lakh ha, permanent pasture, grazing land, trees 9.04 lakh ha and groves 2.77 lakh ha and fallow land including current 15.82 lakh ha and others 5.23 lakh ha. The cropping intensity is 121.93% (Anon., 2017).

Distribution of operational holdings of land depicts a skewed pattern in the state, small and marginal holdings accounted for 75.20 per cent of the total number of holdings while occupying only 35.30 per cent of the total area. In other words, a large proportion of farmers in the state possessed a small proportion of land and vice versa (Karnataka Agriculture Census, 2011).

In overall Karnataka state is self-sufficient in foodgrains production. However, the production is fluctuating over the years due to frequent droughts. Food grains production in the State increased from 110 lakh tonnes in 2009-10 to 155.6 lakh tonnes in 2015-16.

To provide agriculture extension service to farmers and encourage them to adopt modern technologies, there are 745 Raita Samparka Kendras (RSKs) established at hubli level. The farmers visit them and avail required farm information, inputs and other supporting services and benefits.

1.2. Human Resource Development

Education is having both intrinsic and instrumental value and provides useful services for the development of nation and enrichment of an individual's life. The positive externalities of education make it a public good, and therefore, it needs to be supplied by the State as markets cannot provide it to the socially optimum allevels. The human development paradigm recognises the role of education in the expansion of choices for well-being, security and comfort. Therefore, the right to education is recognized as one of the fundamental human rights and, the drive towards universal elementary education aims at ensuring its delivery.

The impressive development of education is termed as education explosion. But unfortunately the quantitative change has not been accompanied by conceptual and qualitative changes of comparable scale and depth, appropriate to the new situations, requirements and needs (UNESCO, 1998). Thus, changes in the field of education have become indispensable for the better. Education should nurture the all-round development of an individual's personality, and thus, make them an active member of knowledge society and a productive citizen of the country.

Karnataka, the fast growing economy's growth is based on the knowledge base of the society. The State has embarked on significant reforms in the education sector with increased public investment to ensure access, equity and quality in education, with community involvement. Hence, there are significant developments in all fields of education. Quality human resources play an important role in development of state or region. One of the measures to assess quality of human resource is the Human Development Index (HDI). It is observed that the level of Human Development Index is much higher in Karnataka (0.615) than at the all-India level (0.609). Among the major Indian states, it ranks ninth, with Kerala occupying the first place. At the international level, Karnataka's position is at 120 while India is at 127. The attainment of human development in Karnataka is more or less on par with that of Egypt and considerably above the level of Pakistan, Nepal, Bhutan and Bangladesh. It can, thus, be argued that the state is well-placed in the context of human development in South Asia. The HDI for the state has increased from 0.432 in 2000 to 0.615 in 2015, showing a 29.75 per cent improvement.

1.3. Higher Education in Karnataka

There has been increasing trend in literacy rates in the state, as it is evidenced the census reports. As per the 2011 census, Karnataka had a literacy rate of 75.60%, with 82.85% of males and 68.13% of females in the state being literate. In 2001, the literacy rate of the state were 67.04%, with 76.29% of males and 57.45% of females being literate.

The Department of Collegiate Education was set up in the year 1960, and has since been striving to make quality higher education affordable and accessible to all sections of students. There are more than 652 degree colleges affiliated with any one of the 25 universities in the state. During 1998, all the engineering colleges in the state were brought under the Vishveshvaraiah Technological University, Belgaum. Similarly, all the medical colleges and paramedical colleges were brought under the jurisdiction of the Rajiv Gandhi University of Health Sciences, Bangalore. In additional, some of colleges were accredited with the status of a Deemed University.

Table 1.1: Growth of Public Universities in Karnataka

S.No.	State Universities	Year of Est.
1	Mysore University, Mysore-570 005	1916
2	Karnataka University, Dharwad-580 003	1949
3	Bangalore University, Bangalore-560 056	1964
4	University of Agricultural Sciences, Bangalore-560 065	1964
5	Gulbarga University, Gulbarga-585 106	1980
6	Mangalore University, Mangalore-574 199	1980
7	University of Agricultural Sciences, Dharwad-580 005	1986
8	Kuvempu University, Shankaraghatta-577 451	1987

9	Kannada University, Hampi, Bellary district, Kamalapura-583 276	1992
10	National Law School of India University, Bangalore-560 072	1992
11	Rajiv Gandhi University of Health Sciences, Bangalore-560 041	1994
12	Karnataka Open University, Mysore-570 006	1996
13	Vishveshwaraiah Technological University, Belgaum-590 010	1999
14	Karnataka State Women University, Bijapur-586 101 (Karnataka)	2004
15	Karnataka Veterinary, Animal & Fisheries Science University, Nandinagar, PB No. 6, Bidar-585 401	2004
16	Tumkur University, 1st Floor, Dr. B.R. Ambedkar Bhavan, M.G. Road, Tumkur- 572 101	2004
17	Davangere University, Shivagangothri, Davangere - 577 002, Karnataka	2009
18	Karnataka State Law University, Hubli	2009
19	Rani Channamma University, Vidyasangama, PB National Highway-04, Bhootramatti, Belagavi-591 156, Karnataka	2010
20	University of Horticultural Sciences, Bagalkot, Udayanagiri, Near Seemikeri, Cross, Bagalkot-587104, Karnataka	2010
21	Vijayanagara Srikrishnadevaraya University, Jnana Sagara Campus, Vinayaka Nagar, Cantonment, Bellary-583 104, Karnataka	2010
22	KSGH Music and Performing Arts University, LJB Road, Near Ashoka Circle, Lakshmipuram, Mysore-570 004, Karnataka	2009
23	Karnataka Folklore University, Gotagodi, Shiggaon Taluk, Haveri district, Karnataka-581197	2011
24	University of Agricultural Sciences, Raichur-584104	2009
25	University of Agricultural and Horticultural Sciences, SH 57, Navule, Shivamogga, Karnataka-577216	2013

As on 2016, there are 221 engineering, 55 Medical Colleges (MBBS degree) in the state which have a summated intake capacity of 8125 students and of them, 16 colleges are run by the Government along with 8 other Super Specialty Institutions. The Government's presence is visible in nominal strength in other system. Relatively, the turnout of Ayurvedic doctors is quite significant. There are 2770 Doctors graduate from 39 Dental Colleges. Ayurvedic, Homeopathic and Unani Doctors serve rural India in larger proportions. 248 Nursing Colleges recognized by Indian Nursing Council turn out 13,840 Nurses (B.Sc. Graduates) every year. Karnataka functions as a nursery for nursing services not only for the State but also for other regions of India, Gulf countries and Europe.

Udupi, Sringeri, Gokarna and Melkote are well-known places of Sanskrit and Vedic learning centres. In addition, there are specialized universities such as Women's University at Vijapura and Kannada University at Hampi apart from three agricultural universities, one horticulture, one agri-horticulture university and one veterinary university in the state.

The state is home to some of the premier educational and research institutions of India such as the Indian Institute of Science, the Indian Institute of Management, the National Institute of Technology Karnataka and the National Law School of India University and IIT, Dharwad.

1.4. Agricultural Education in Karnataka

Agriculture Education began in Karnataka as a small agricultural research farm in 1899 on 30 acres of land donated by Her Excellency Maharani Kempa Nanjammanni Vani Vilasa Sannidhiyavaru, the Regent of Mysore and appointed Dr. Lehmann, a German Scientist to initiate research on soil crop response with a laboratory in the Directorate of Agriculture. Later under the initiative of the Dewan of Mysore, Sir M. Vishweshwaraiah, the Mysore Agriculture Residential School was established in 1913 at Hebbal, which offered Licentiate in Agriculture and later offered a diploma programme in agriculture during 1920. The school was upgraded to Agriculture College in 1946 which offered four years degree programs in Agriculture. The Government of Mysore headed by Sri. S. Nijalingappa, the then Chief Minister, established the University of Agricultural Sciences on the pattern of Land Grant College system of USA and the University of Agricultural Sciences Act No. 22 was passed in Legislative Assembly in 1963. Dr. Zakir Hussain, the Vice President of India inaugurated the University on 21st August, 1964 in Bangalore.

At present, there are three Agriculture, one Agriculture and Horticulture, one Horticulture and one Veterinary Universities are functioning in the state. University of Agricultural Sciences, Bengaluru is the first Agriculture University in Karnataka, which was established during May 25, 1963 with having Agriculture, Horticulture and Veterinary Colleges at different parts of Karnataka. Later University of Agricultural Sciences, Dharwad was established on October 1, 1986 with five colleges. On February 23, 2004, Karnataka Veterinary, Animal and Fisheries Sciences University, Bidar was established with seven colleges which separated from UAS, Bangalore, UAS, Dharwad. University of Horticultural Sciences, Bagalkot was established with jurisdiction of entire state during November 22, 2008 with all horticultural colleges separated from Agriculture Universities. Afterwards one more Agriculture University as UAS, Raichur was established in the year 2009 with four colleges and having jurisdiction of Hyderabad, Karnataka region. Recently University of Agricultural and Horticultural Sciences, Shimoga was established in September 21, 2012 with four (one Agriculture, one Forestry and two Horticulture) colleges (Fig. 1.1).

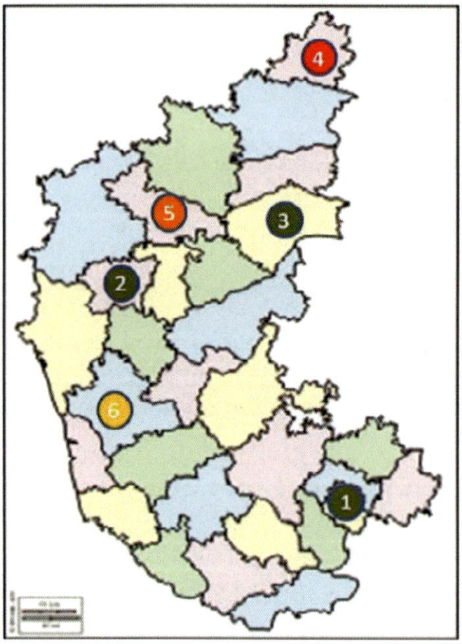

INDEX

1. University of Agricultural Sciences, Bangaluru (May 25, 1963)
2. University of Agricultural Sciences, Dharwad (October 1, 1986)
3. University of Agricultural Sciences, Raichur (April 1, 2009)
4. Karnataka Veterinary, Animal and Fisheries Sciences University, Bihar (February 23, 2004)
5. University of Horticultural Sciences, Bagalkot (November 22, 2008)
6. University of Agricultural and Horticultural Sciences, Shimoga (September 21, 2012)

Fig. 1.1: Distribution of Farm Universities in Karnataka

1.4.1. Women in Agriculture Education

Women are the backbone of the development of rural and national economies. Women guarantee livelihoods, especially in rural areas. As a result of their great efforts in agricultural production and also helps to guarantee their self-sustenance. It is also noteworthy that more females are taking admissions in agriculture and allied courses in the universities. The data in Table 1.2 presents the trends of female students in agriculture education; the data reveals increasing number of female students in agriculture; the proportion of female students has increased from 28.78 per cent in 2011 to about 34.87 per cent in 2015. Thus, over the years the proportion of trained female human resource in agriculture has increased (Table 1.2). It may also due to better employment opportunities in public sector organisations, as there is a 30 per cent reservation for women.

Table 1.2 Female Students in Agriculture Education in Karnataka

Year	Per cent Female Students (Diploma)	Per cent Female Students (UG)	Per cent Female Students (PG)	Per cent Female Students (Ph.D.)	Per cent Female Students (Total)
2010-11	20.67	37.44	25.18	31.82	**28.78**
2011-12	18.94	37.88	31.43	30.89	**29.79**
2012-13	21.05	39.16	35.05	36.60	**32.97**
2013-14	23.11	41.26	36.93	32.91	**33.55**
2014-15	25.94	40.97	34.80	37.78	**34.87**

The quality of agriculture education is the key for solving multifaceted problems agriculture and allied sectors. Thus, there is an urgent need to quantitatively and qualitatively assess the current stock size and status of the human resource in agriculture sector with special reference to requirement of personnel for various sub-sectors of agriculture and allied activities. Also, it is imperative to assess the future human resource needs in consonance with the targeted growth rates of the various sub-sectors of agriculture.

In this context, forecasting of future agricultural human resources requirements assumes considerable significance in identifying the trends and visualizing the emerging technologies and the corresponding needs for the appropriate human resources mix. The assessment of existing human resources, the sectoral growth, capacity utilisation of qualified graduates is indispensable to envisage the future agricultural education needs.

1.5 Need for Assessment of Human Resource in Agriculture and Allied Sectors

Trained human resources are needed in different sectors of agriculture for targeted growth. The sectors are to be serviced by human resources with higher skills than before to ensure technology generation, its transfer to and more importantly its application at the grassroot level. Further, agricultural human resources with diverse skills is today required by a wide-ranging and fast expanding food processing industry, corporate and unorganized sector.

In this context, forecasting of future agricultural human resources requirements assumes considerable significance in identifying the trends and visualizing the emerging technologies and the corresponding needs for the appropriate human resources mix. The assessment of existing human resources, the sectoral growth, capacity utilisation of qualified graduates is indispensable to envisage the future agricultural education needs. Considerable efforts are needed to evolve alternative strategies to cope with the scenario outlined above including intensive problem-oriented research, transfer of solutions to the masses and skill development as well as adequate skill up gradation.

The present study is conducted to address the developmental needs of the economy with special reference to education in agriculture and its allied sectors which have an important place in the state economy due to their role and potential in providing employment to skilled, semi skilled as well as unskilled persons. Skill needs are changing and hence, a need for analyzing the skill gaps, the institutional mechanism in imparting education and assess the future requirements of the sector in the context of human resources.

1.6 The Objectives of the Study are:

- To estimate the demand for services of agricultural graduates in public and private sector.
- To assess the employment opportunities for agricultural graduates turning out in the state.
- To review ICAR and SAU schemes for HRD of students in SAU for improvising.

- To identify gaps in quantity and quality of supply of skilled man power in agriculture.
- To suggest measures and strategies for the development of skilled human resources in agriculture and allied sectors.

❑❑❑

Chapter 2

Performance of Agriculture and Allied Sectors

The main features and performance of agriculture and allied sectors *viz.*, horticulture, veterinary and animal husbandry, dairy, fisheries, forestry, agriculture engineering, home science, sericulture and agricultural biotechnology in Karnataka has been discussed in this chapter.

2.1 Agriculture and National Economy

The history of Indian agriculture dates back to 10,000 years. Indian agriculture began during 9000 BC as a result of early cultivation of plants and domestication of crops and animals. The middle ages in India saw irrigation channels that reached a new level of sophistication. Land and water management systems were developed with the objective of providing uniform growth. Presently, the agricultural sector employed 55 per cent of the total workforce in India, and despite a steady decline of its share in the GDP, it still remains the largest economic sector. Agriculture plays a significant role in the overall socio-economic development of India.

According to National Accounts Statistics, contribution of agriculture to the GDP has been steadily declining over the years; from around 56 per cent in 1950-51, it has fallen to 34 per cent in 1990-91, 17.01 per cent in 2014-15. In spite of this decline, the sector continues to be the pre-dominant occupation of about 55 per cent of working population in the country, although this proportion is coming down slowly.

India ranks second worldwide in farm output. Agriculture and allied sectors like forestry, logging and fishing accounted for 17% of the GDP and employed 55 % of the total workforce in 2014. As the Indian economy has diversified and grown, agriculture's contribution to GDP has steadily declined from 1951 to 2016, yet it is still the largest employment source and a significant sector of the overall socio-economic development of India. Crop yield per unit area of all crops has increased since 1950, due to the special emphasis placed on agriculture in the five-year plans and steady improvements in irrigation, technology, application of modern agricultural practices and provision of agricultural credit and subsidies since the Green Revolution in India. However, international comparisons reveal the average yield in India is lower to the extent of 30% to 50% of the highest average yield in the world. The states of Uttar Pradesh, Punjab, Haryana, Madhya Pradesh, Andhra Pradesh, Telangana, Bihar, West Bengal, Gujarat, Maharashtra and Karnataka are key contributors to Indian agriculture.

India receives an average annual rainfall of 1,208 mm and a total annual precipitation of 4000 billion cubic metres, with the total utilisable water resources, including surface and groundwater, amounting to 1123 billion cubic metres. 546,820 square kilometres (211,130 sq m) of the land area, or about 39% of the total cultivated area, is irrigated.

India's inland water resources including rivers, canals, ponds and lakes and marine resources comprising the east and west coasts of the Indian ocean and other gulfs and bays provide employment to nearly six million people in the fisheries sector. In 2010, India had the world's sixth largest fishing industry. India is the largest producer in the world of milk, jute and pulses, and also has the world's second largest cattle population with 170 million animals in 2011. It is the second largest producer of rice, wheat, sugarcane, cotton and groundnuts as well as the second largest fruit and vegetable producer, accounting for 10.9% and 8.6% of the world fruit and vegetable production respectively. India is also the second largest producer and the largest consumer of silk in the world, producing 77,000 tonnes in 2005. India is the largest exporter of cashew kernels and cashewnut shell liquid (CNSL). Foreign exchange earned by the country through the export of cashew kernels during 2011-12 reached Rs.4,390 crore (Cashew Export Promotion Council of India, CEPCI). 131,000 tonnes of kernels were exported during 2011-12. There are about 600 cashew processing units in Kollam, Kerala. India's food grains production remained stagnant at approximately 252 million tonnes (mt) during both the 2015-16 and 2014-15 crop years (July- June). India exports several agriculture products, such as Basmati rice, wheat, cereals, spices, fresh fruits, dry fruits, buffalo beef meat, cotton, tea, coffee and other cash crops particularly to the Middle East, South-east and East Asian countries.

2.2 Agriculture in Karnataka

The scientific agriculture begins in Karnataka by way back 1836 by Sir Mark Cubbon (Commissioner, 1834-1861) took steps to set up the agriculture society in Mysuru. In 1857, Visternew was appointed as Superintendent, Bengaluru Botanical Garden (Lalbaugh). He took various measures to improve horticultural activities. In 1862, Signorde Vichi introduced new mulberry varieties, silkworm strains and added new dimensions in sericulture in Mysuru. In 1869, Mr.Von Somerin, as Conservator of Forests gave the beginning to scientific forestry in Mysuru.

It is our bounded duty to remember with gratitude Rajarishi Sri Nalwadi Krishnaraja Wadeyar, the then Maharaja of Mysuru (1894-1940). He, having realized potentials of Agriculture, took up various developmental activities systematically and put Karnataka first in India, in various aspects of agriculture. In 1899, he appointed Dr. Lehman (a German chemist) who set up a Soils Laboratory and the multi-disciplinary agricultural laboratory in 1901. Dr.Lesile C.Coleman (a Canadian Scientist) was also appointed in 1905 as Scientist to take up entomology and pathology research. In 1907, Coleman took over the charge of Lehman. During 1912, a separate Department of Agriculture was started bifurcating from the Revenue and other Departments based on the recommendations of Dr.Leslie Coleman, who was also appointed as first Director of Agriculture in July 5, 1913. He takes the credit of starting the four years Diploma Course in Agriculture at Hebbal, Bengaluru on 5th July, 1913, being the first in Asia. Later in 1916, he started agricultural school at Chikkanahalli (Tumkuru) and later at Anekal (Bengaluru), Somanahalli (Mandya), etc. He also started several research stations, at Thirthalli (Areca nut), Balehonnur (Coffee), V.C.Farm, Mandya (Sugarcane, Paddy, Ragi) and Babbur (Oil seeds), to cater to the needs of local agricultural problems. The research station at Mandya was probably the first centre in India, established for irrigated ecosystem in 1930. Dr.Coleman prepared a master plan to start agriculture degree college before he retired in 1937. He visited Mysuru state again in 1950 at the invitation of the then Mysuru Maharaja Sri Jayachamarajendra Wodeyar and prepared comprehensive agriculture developmental plan for the state. In 1946, agricultural colleges were started at Hebbal and Dharwad, which later graduated into University of Agriculture Sciences in 1966.

As much as the farmers of Karnataka, owe a debt of gratitude to Dr.Coleman for his vision and determination for the overall development of agriculture education, research and development of Karnataka.

Agriculture is the mother of all culture and the progenitor of human civilization. The recorded history provides ample proof for rich heritage of Indian agriculture. The agriculture that flourished during Aryan and the Indus Valley Civilization dating back to 3000 B.C., stands testimony to Indian culture through agriculture. Agriculture provides the main economic support to the state. The importance had been given to agriculture in Karnataka, was indicated by the statement of famous Kannada poet Sarvajanya who says that meti vidye (agricultural science) is superior to koti vidye (millions of other sciences).

As 66% of the cultivated area in the State is under rainfed cultivation and failure of rains for the past consecutive four years (2012 to 2016) has made a severe impact on the agricultural production in the state.

2.3 Agriculture and State Economy

Agriculture plays an important role in the overall growth of Karnataka's economy despite a fall in its share in the state domestic product. Agricultural growth of 4% and above is a necessary condition for inclusive growth in the state. It is also important, as it brings in price stability, food security and income and livelihood of the majority. For about 70 per cent rural residents of Karnataka, agriculture is the major occupation. A total of 123,100 km^2 of land is cultivated in Karnataka, constituting 25.3% of the total geographical area of the state. According to the 2011 census, farmers and agricultural labourers formed 55% of the workforce of Karnataka. Agriculture in Karnataka is heavily dependent on the south-west monsoon since the extent of arid land in the state is second only to Rajasthan. Only 26.5% of sown area (30,900 km^2) is subjected to irrigation. The state has three agricultural seasons – Kharif (April to September), Rabi (October to December) and Summer (January to March). The main crops grown here are rice, ragi, jowar, maize, and pulses (Tur and Gram) besides oilseeds and number of cash crops. Cashews, coconut, arecanut, cardamom, chillies, cotton, sugarcane and tobacco are among the other crops produced in the state. Karnataka is the largest producer of coarse cereals, coffee, raw silk and tomatoes among the states in India. A majority of the 35 billion silk industries in India is co-centred in Karnataka state, particularly in Mysore and Bangalore Rural districts.

In Karnataka, horticulture crops occupy about 15.21 lakh hectares with an annual production of about 96.60 lakh tonnes. The income generated from horticulture constitutes over 40% of income generated from agriculture and allied sectors. In floriculture, Karnataka occupies the second position in India, in terms of production and 700 tonnes of flowers (worth Rs.500 million) were produced in 2014–15. Karnataka is highly progressive with regard to vegetable production, and enjoys this advantage because of favourable climatic conditions without any extremes in temperature. The fisheries sector is now emerging as one of the most important in allied agriculture activities in the state.

Agricultural growth rate (current prices) has fluctuated from a low to high 15.87 % to 16.63 %) from (2004-05 to 2005-06), during 2005-06 to 2006-07 from high to low of 14.64 per cent and again down to 14.42 per cent in 2007-08, same trend was continued upto 2012-13 (13.55%). Hence, from 2013-14 onwards increasing trend was observed 14.27 in 2013-14 and 14.75 in 2014-15 as shown in (Table 2.1). Although per capita income has shown considerable increase owing to the growth in manufacturing and service sectors, the per capita GSDP in rural areas from agricultural activities remains low and hence, the inter- sectoral inequalities are growing sharply. A large number of workers depend on agriculture relative to the output it generates, resulting in lower labour productivity compared to non-agricultural sectors. Less than one-fifth of the GSDP comes from more than 50 per cent of the workforce in the state. This has serious implications for improving the welfare of rural population and alleviation of poverty.

The Gross State Domestic Product (GSDP), which was Rs. 166747 crore during 2004-05, rose to Rs. 337555 crore in 2009-10 and touched Rs. 702131 crore (current prices) in 2014-15 (Table 2.1 and Fig. 2.1). The per capita income (in current prices) increased from Rs. 30,138 in 2004-05 to Rs. 1,14,056 during 2014-15. Agriculture took a share of 14.75 per cent of the State GSDP in 2014-15 as against 15.87 per cent in 2004-05. Surprisingly, the share of industrial sector also indicated a declining trend over the years and its contribution declined from 30.28 per cent in 2004-05 to 23.27 per cent of the State GSDP in 2014-15.

Table 2.1: Sectoral shares of GSDP for Karnataka (Current Prices)

Sectors	Agriculture (%)	Forestry & Logging (%)	Fishing (%)	Industry (%)	Services (%)	Total GSDP (Rs. in crore)	Per Capita GSDP (Rs.)
2004-05	15.87	2.46	0.38	30.28	51.02	166747	30138
2005-06	16.63	2.49	0.43	29.63	50.82	195904	34988
2006-07	14.64	2.28	0.41	31.99	50.67	227238	40115
2007-08	14.42	2.28	0.34	31.81	51.15	270628	47237
2008-09	13.09	2.3	0.48	31.34	52.8	310312	53569
2009-10	13.86	1.81	0.52	29.1	54.72	337555	57650
2010-11	15.20	2.15	0.6	28.66	53.38	410703	69411
2011-12	13.62	1.9	0.6	28.36	55.52	455213	76148
2012-13	13.55	2.26	0.59	26.34	57.25	522673	86561
2013-14	14.27	2.18	0.62	24.68	58.24	614607	100797
2014-15	14.75	2.17	0.61	23.37	59.1	702131	114056

Source: GoK, Economic Survey, 2016-17

The sectoral growth rates during 2004-05 to 2014-15 period for agriculture and allied activities, industry and services are 4.69 per cent, 5.49 per cent and 8.70 per cent, respectively. In general, the trends in the annual growth rate of GSDP correspond to the growth rates of industry and service sectors, and the State GSDP grew at 7.12 per cent per annum between 2004-05 and 2014-15. At the same time, per capita GSDP has increased from Rs. 30138 to Rs. 114056 in 2004-05 to 2014-15 (Table 2.1).

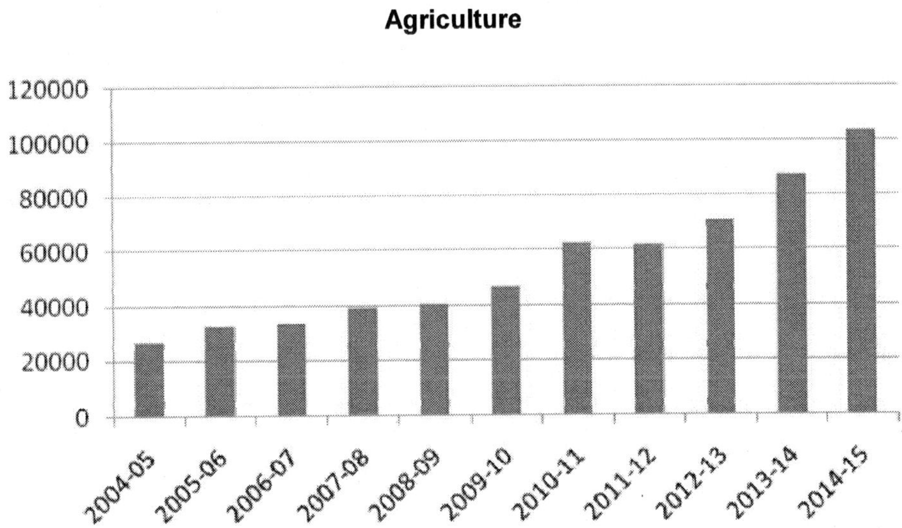

Fig. 2.1: Recent Trends in Agriculture GSDP

2.4 Performance of Agriculture Sector

Some important features that have an impact on and are of grave concern to agriculture today, are the problems of small and marginal farmers, low productivity, issues relating to transfer of technology against the background of almost collapse of public extension services, the regional disparities in modernization of agriculture, constraints on land availability, depleting groundwater and deterioration of soil health, changing patterns of monsoon, climate change, commercialization of agriculture, movement of labour force away from agriculture, feminization of agriculture, post-harvest losses, policies of globalization and liberalisation and so on.

2.4.1 Decline in Soil fertility

There are varied types of soils in Karnataka. Six broad groups of soil orders are recognized, based on differences in soil formation processes, as reflected in the nature and sequence of soil horizons. Black soils are found in northern Karnataka, whereas red and red loamy soils are prominent in southern Karnataka. Laterite soils are found in malnad and coastal areas of the state. Soil fertility in the state has been declined in trend, in over a period of time due to intensive cultivation and chemicalisation of agriculture. This has serious implications on productivity growth of crops in the state.

2.4.2 Change in Land-use Pattern

Land is a finite resource and its demand for diverse purposes is increasing day-by-day. There are large areas of utilizable but wasteland resources. Moreover, the pressure of commercialization is increasingly forcing out productive land resources for non-agricultural uses. The analysis of land-use data from 1966 to 2015 indicates structural changes in land-use pattern. Areas under permanent pasture decreased consistently and areas under non-agricultural uses increased sharply during the past five decades. Fallow land, which was around 11 lakh ha, increased to 13.50 lakh ha after mid-nineties and 15.82 lakh ha in 2014-15, the percentage reached above 10 per cent of the total (190.50 lakh ha) geographical area of the state. This was mainly because of deficient rainfall and the inability of the government to provide irrigation facilities in those years. The net sown area also started shrinking after 1990s. The net sown area, which accounted for 104.01 lakh ha during 1988, declined to 100.44 lakh ha during 2014-15 (Table 2.2).

Table 2.2: Trends in Area under various Land-use categories in Karnataka

Area in lakh hectares

Land-use category	1967-68	1977-78	1987-88	1997-98	2012-13	2013-14	2014-15
Non-Agr. Purposes	8.76	10.36	11.72	12.88	14.36	14.44	14.61
Permanent Pastures	16.76	14.49	11.32	10.03	9.08	9.06	9.04
Current Fallows	10.65	13.05	10.90	13.59	18.22	16.70	15.82
Net Sown Area	100.67	99.40	106.21	104.01	97.93	99.23	100.44
Gross Irrigated Area	12.97	17.16	23.83	29.70	32.15	35.23	56.31

Source: GOK, Dept Agri. Annual report, 2014-15

2.4.3 Decreased Land holding Size

It is interesting to note that the number of small and marginal holdings as well as their share in the total operated area is increasing over the years. The increase in small and marginal holdings and area operated became more conspicuous after 2000-01 (Table 2.3). Small and marginal farmers (operating < 2 ha) account for 76 per cent of the holdings and share roughly 37 per cent of the operated area in Karnataka. The average size of operated area of all the land size classes is declining. This clearly indicates the increasing fragmentation of land holdings in the state.

Table 2.3: Trends in number of operational holdings according to major size classes of Agricultural Census1, 1970-71 to 2010-11.

Size Class	1970-71	1976-77	1980-81	1985-86	1990-91	1995-96	2000-01	2005-06	2010-11
Marginal	1081	1274	1489	1792	2262	2610	3252	3655	3849
Changes (%)		17.8	16.9	20.3	26.2	15.4	24.6	12.4	5.3
Small	840	888	1057	1293	1586	1707	1909	2014	2138
Changes (%)		5.8	19	22.3	22.7	7.6	11.8	5.4	6.2
Semi-medium	788	818	918	1035	1163	1204	1259	1278	1267
Changes (%)		3.8	12.2	12.7	12.4	3.5	4.6	1.5	-0.9
Medium	623	632	662	646	636	594	569	555	511
Changes (%)		1.4	4.7	-2.4	-1.5	-6.6	-4.2	-2.6	-7.9
Large	219	199	183	153	129	106	90	79	67
Changes (%)		-9.1	-8	-16.4	-15.7	-17.8	-15.1	-11.1	-15.2
All Size	3551	3811	4309	4919	5776	6221	7079	7581	7832
Changes (%)		7.3	13.1	14.1	17.4	7.7	13.8	7.1	3.3

Source: Karnataka Agriculture census, 2011

The shrinking sizes of holdings and the high proportion of unviable farmers impinge upon the quality of life. The economic non-viability of a large number of small and marginal holdings and the hardship faced by their holders are reflected in the growing number of suicides committed by farmers. This calls for initiating a new phase of land reforms, especially policies related to land tenancy.

2.4.4 Change in Population and Workforce Composition

The population of Karnataka, which was 5.29 crore in 2001, has increased to 6.11 crore in 2011. The female population of Karnataka which accounted for 49.10 per cent of the total population in the state during the 2001 census, has gone up marginally to 49.19 per cent in the 2011 census. The total population growth in this decade was 15.67 per cent while in the previous decade, it was 17.25 per cent. The population of Karnataka forms 5.05 per cent of the all-India population in 2011 as against 5.14 per cent during the 2001 census. Rural population, which formed 66 per cent of the total population of Karnataka in 2001, declined to 61.43 per cent in 2011 (Table 2.4). The proportion of rural female population declined from 49.41 per cent in 2001 to 49.19 per cent in 2011. On the contrary, the proportion of female population in urban Karnataka rose marginally from 48.50 per cent in 2001 to 48.91 during 2011.

Table 2.4: Population of Karnataka during 2001 and 2011 Census

Population in '000'

Population	Census, 2011			Census, 2001		
	Total	Rural	Urban	Total	Rural	Urban
Persons	61,095	37,469	23,626	52,851	34,889	17,962
Male	30,967	18,929	12,037	26,899	17,649	9,25012
Female	30,129	18,540	11,589	25,952	17,240	8,712

Source: Karnataka Census, 2011

According to the 2001 census, workers constituted 44.53 per cent of the total population and the rest 55.47 per cent were non-workers. There were 2.35 crore main workers accounting for roughly 37 per cent of the total population as per the 2001 census. The cultivators (29.25 per cent) and agricultural labourers (26.46 per cent) formed 52 per cent of the workforce of Karnataka (Fig.2.2). Workers in the household industry accounted for 4 per cent, whereas other workers constituted 40.21 per cent of the total workers in Karnataka in 2001. As per the 2011 census, 2.79 crore people accounting for 45.62 per cent of the population form the workforce and among these, cultivators accounted for 21.66 per cent and agriculture labour constituted 18.37 per cent in the state. It is interesting to note that the share of both cultivators as well as agricultural workers has declined in 2011 from the 2001 census figures.

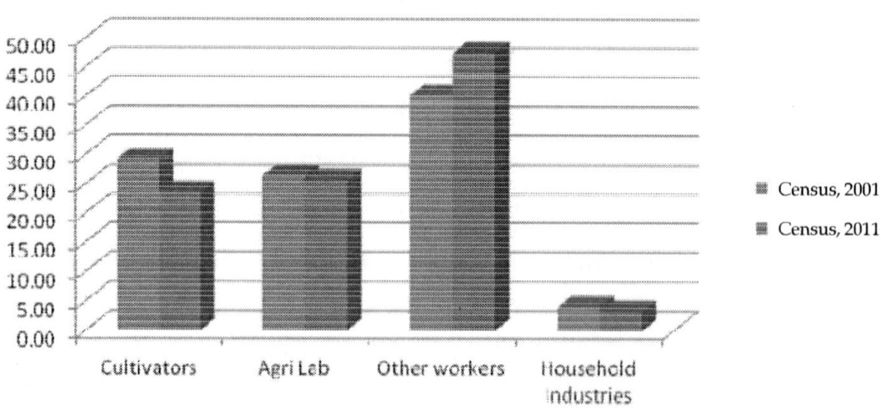

Fig. 2.2: Changing Proportion of Workers in Karnataka

2.4.5 Trends in Area, Production and yield of Agriculture Crops

The cropping pattern of the region is influenced not only by agro-climatic conditions like rainfall, soil, temperature, *etc.*, but also by government policies and programmes for crop production in the form of subsidies, support prices, tariffs and speed of infrastructure development. The overall trends in area allotted for various crops during five decades, show that cropping pattern in Karnataka is dominated by food crops, with a share of more than 60 per cent of the gross cropped area in the state. Rice, sorghum and finger millet were the major cereals till 2000-01. However, the share of maize crop went up substantially after 2005 due to improved productivity and prices. The compound annual growth of agriculture crops area (0.21 %), production (3.35 %) and productivity (3.09 %) from 2000-01 to 2014-15 (Table 2.5 and Fig. 2.3).

Table 2.5: Growth in agriculture crops (area, production and productivity) in Karnataka during 2000 to 2015

Year	Area (Lakh ha)	Production (Lakh tonnes)	Yield (kg/ha)
2000-01	78.04	109.6	1478
2001-02	71.75	87.0	1276
2002-03	70.24	66.6	999
2003-04	68.82	65.6	1004
2004-05	75.67	104.9	1459
2005-06	75.96	115.4	1598
2006-07	73.94	93.3	1328
2007-08	78.73	120.5	1611
2008-09	74.61	111.3	1571
2009-10	79.56	108.0	1340
2010-11	82.37	137.9	1762
2011-12	74.14	117.7	1667
2012-13	73.00	109.6	1581
2013-14	75.46	133.1	1857
2014-15	71.93	126.2	1847
CAGR (%)	**0.21**	**3.35**	**3.09**

Source: Agriculture Dept. GOK, Annual Report, 214-15

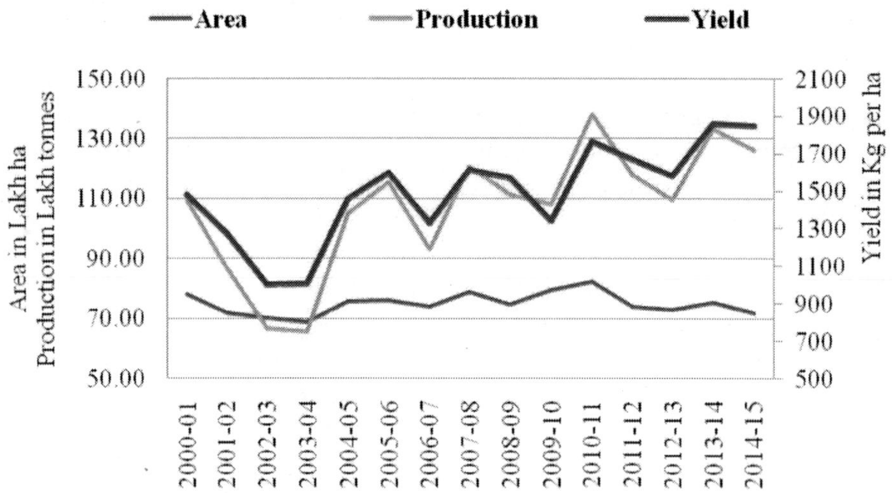

Fig. 2.3: Growth in area, production and productivity of agriculture crops

2.4.5.1 Area, Production and yield of Food grains

The area under food grains (cereal and pulses) has been reduced from 78 lakh ha in 2000 to 71 lakh ha in 2015. Production and yield of food grains during 2000 to 2015, and their growth rates are given in Table 2.6. The food grain production has increased from 109.6 lakh tonnes in 2000-01 to 126.2 lakh tonnes in 2014-15. The productivity of food grains increased from about 1478 kg/ha to 1847 kg/ha (from 2000-01 to 2014-15) with growth rate 3.09 per cent. Similar trend was observed in oilseeds and commercial crops.

Table 2.6: Growth rate of area, production and yield of food grains in Karnataka

Year	Area (lakh ha)	Production (lakh tonnes)	Yield (Kg/ha)
2000-01	78.04	109.6	1478
2001-02	71.75	87.0	1276
2002-03	70.24	66.6	999
2003-04	68.82	65.6	1004
2004-05	75.67	104.9	1459
2005-06	75.96	115.4	1598
2006-07	73.94	93.3	1328
2007-08	78.73	120.5	1611
2008-09	74.61	111.3	1571
2009-10	79.56	108.0	1340

2010-11	82.37	137.9	1762
2011-12	74.14	117.7	1667
2012-13	73.00	109.6	1581
2013-14	75.46	133.1	1857
2014-15	71.93	126.2	1847
CAGR (%)	**0.21**	**3.35**	**3.09**

Source: Annual Report, Dept. of Agriculture, Govt. of Karnataka, (2015-16)

2.4.6 Input Use and Other Services

The use of inputs such as high-yielding crop varieties, chemical fertilizers, plant protection chemicals as well as farm machinery in agriculture has increased over the years, which facilitated improvement in productivity and resulted in increased crop production.

2.4.6.1 Area under High-Yielding Varieties

The area under high-yielding varieties (HYVs) of major crops increased from 30.18 lakh hectares in 2000-01 to 41.44 lakh ha in 2009-10 and further to 44.10 lakh ha in 2014-15 (Table 2.7). Area under HYVs grew at a faster rate between 2000-01 and 2005-06 than during 2005-06 to 2009-10 and 2009-10 to 2014-15. The area under HYV maize and wheat increased by more than 35 per cent whereas the area under HYV rice and sorghum recorded a modest increase of 2 and 3 per cent, respectively during the five-year period from 2009-10 to 2014-15.

Table 2.7: Area under HYV under Major Crops in Karnataka (Lakh ha)

Crops	2000-01	2005-06	2009-10	2014-15
Rice	12.30	13.11	13.28	14.16
Sorghum	7.12	10.62	10.88	11.34
Pearl millet	4.19	4.15	2.93	2.86
Maize	6.57	9.19	12.32	13.49
Wheat	0	1.47	2.03	2.25
All	30.18	38.54	41.44	44.10

Source: Agriculture Dept. GOK, Annual Report, 214-15

2.4.6.2 Changing Fertilizer Consumption

Expansion in irrigated area and coverage of high-yielding varieties resulted in higher demand and use of chemical fertilizers in Karnataka. Total fertilizer consumption increased to 18.31 lakh tonnes by 2014-15 as against 12.5 lakh tonnes during 2001-02 (Table 2.8). The highest fertilizer consumption was noticed 22.1 lakh tonnes during 2011-12 thereby decreasing trend was observed due to shift of farmers towards organic farming and stop of indiscriminate use of fertilizers. Consumption of phosphatic and potassic fertilizers was relatively higher, when compared to nitrogenous fertilizers. The use of fertilizers increased from 103 kg per hectare in 2001-02 to 194 kg per ha during 2014-15.

Table 2.8: Total and Per Hectare Consumption of Fertilizers in Karnataka

Year	Fertilizers used per hectare (Kg)				Nutrient-wise consumption (Lakh tomes)			
	N	P	K	Total	N	P	K	Total
1991-92	30	24	13	67	4.50	2.90	1.70	9.10
2001-02	52	30	18	103	6.70	3.60	2.20	12.50
2005-06	58	33	26	117	7.50	4.40	3.40	15.30
2010-11	78	53	30	162	10.20	7.00	4.00	21.10
2011-12	80	55	31	166	11.04	7.87	3.30	22.21
2012-13	84	51	28	163	8.92	3.89	2.49	15.30
2013-14	91	58	33	182	9.63	4.91	3.01	17.55
2014-15	96	61	37	194	10.01	4.98	3.32	18.31

Source: GOK, Statistical Abstract of Karnataka, various issues

2.4.6.3 Pesticides Consumption

Use of plant protection chemicals (insecticides and pesticides) remained more or less stagnant and the technical grade material used hovered around 1821 tonnes (except in 2006-07) despite the expansion in the area covered, which increased from 38 lakh ha in 2005-06 to 84.10 lakh ha in 2014-15 (Table 2.9).

Table 2.9 : Use of Pesticides and Total Area covered in Karnataka

Year	Area covered (Lakh ha)	Technical grade material (Tonnes)
2005-06	38.00	1698
2006-07	24.00	942
2007-08	58.50	1588
2008-09	70.00	1675
2009-10	73.00	1700
2010-11	74.00	1720
2011-12	76.00	1731
2012-13	78.00	1758
2013-14	79.00	1803
2014-15	81.00	1821

Source: GOK, Statistical Abstract of Karnataka, various issues

Awareness about the health hazards of pesticide use, introduction of integrated pest management practices and promotion of organic farming have resulted in reduced use of pesticide per unit area during last five year. This is mainly due to area under Bt cotton has increased.

2.4.6.4 Quality Seeds

Availability and access to quality seeds are key factors in improving productivity and production of crops. Improved as well as hybrid seeds of various crops produced by private and public sector seed companies are available in the state. The state government has undertaken a programme of production and distribution of certified quality seeds from 2007-08. Under the seed replacement norms, seed replacement rates fixed for different agricultural crops during 2014-15 are : cereals 23-34; pulses 23-26%; oilseeds 15-87%; cotton 17% and all hybrids 100%. The production and distribution of certified quality seeds during 2007-08 through 2014-15 is presented in Table 2.10.

Table 2.10: Production and Distribution of Certified quality Seeds in Karnataka

Particulars	Production (qtls)	Distribution (qtls)
2007-08	580950	919860
2008-09	707769	919976
2009-10	1087450	1035123
2010-11	1128258	1292765
2011-12	1145824	1159575
2012-13	1055650	1260990
2013-14	1071450	1280720
2014-15	1088350	1265176

Source: GOK, Economic Survey of Karnataka, 2011-12

2.5.4 Storage, Godowns and Markets

It is estimated that about 6 per cent of the food grains and roughly 30 to 40 per cent of the horticulture produce, mainly fruits and vegetables, are lost due to non-availability of proper storage facilities. This affects the post-harvest management of agricultural and horticultural produce. Lack of storage facilities creates a glut in the market during the harvest season and producers are compelled to sell their produce at lower prices. The storages and godowns are operated by private as well as public sector companies. The Karnataka State Warehousing Corporation has a storage capacity 9.04 lakh MT, of which 4.97 lakh MT is owned and 4.07 MT is hired. Similarly, there are 157 cold storages owned by private, co-operative and public sector units with a capacity of 3.97 lakh MT (GoK, 2015).

Most of the agriculture produce in the state is marketed through the Agriculture Produce Marketing Committee (APMCs) yards. There are 504 regulated markets in the state, of which 151 are main markets and 353 sub-markets. The density of regulated markets for agriculture produce varies across the districts and it ranges from 6 markets in Bangalore rural and Udupi to a maximum of 47 markets in Belgaum. The average total turnover in the regulated markets is around Rs. 22,455 crore per annum.

2.5.5 Farm Machinery and Equipment

Traditionally, farming was carried out using bullock power and human labour. However, increasing cost of maintaining bullock pair, shortage of labour for farm operations and paucity of time induced farmers to adopt mechanization. Mechanization facilitates timely completion of farm operations with desired results and precision. It also helps in reducing the drudgery associated with different conventional operations. A large number of farm equipment and machinery are used in farming. Farm machinery and equipments are distributed to farmers under various state sector and centrally sponsored schemes. Farmers get 50 per cent subsidy for farm machinery and equipments.

2.5 Sericulture Sector

Sericulture is one of the major employment generating sectors in the State and its growth has immense employment generation potential, particularly in rural Karnataka. The area under mulberry cultivation in the State was about 87,597 hectares during 2015-16. The production of cocoons, quantity of cocoons marketed, raw silk production was high in 2015-16, compared to 2014-15 and total employment in industry was high in 2015-16, compared to 2014-15. Details of the State's production and imports of silk yarn are presented in Table 2.11. During 2015-16, the estimated mulberry silk production in Karnataka was 9823 MTs (Table 2.12).

Table 2.11: Sericulture industry in Karnataka, 2012-13 to 2015-16

S.No.	Category	Unit	2012-13	2013-14	2014-15	2015-16
1	Area under mulberry	'000 ha	74.13	80.87	88.489	87.597
2	Production of cocoons	'000 MTs	49.44	61.41	68.759	70.436
3	Quantum of Cocoons marketed	'000 MTs	48.95	45.99	48.492	53.541
4	Raw silk production	'000 MTs	7.06	8.57	9.645	9.823
5	Employment in sericulture	lakhs	9.63	10.51	11.5	11.38

Source: Department of Sericulture

Table 2.12: Production of silk yarn and import of silk yarn in Karnataka, 2000-01 to 2015-16

Year	Karnataka Silk Production (in tonnes)	Import of Silk Yarn to India (in tonnes)
2000 – 2001	8121	4713
2001 – 2002	8200	6808
2002 – 2003	8728	9054
2003 – 2004	6760	9258
2004 – 2005	7302	7185
2005 – 2006	7471	8334
2006 – 2007	7883	5567
2007 - 2008	8240	7921
2008 – 2009	7238	8392
2009 – 2010	7360	7341
2010 – 2011	7338	5870
2011 – 2012	7796	5673
2012 – 2013	7063	4951
2013 – 2014	8574	3259
2014 – 2015	9645	3489
2015 - 2016	9823	3529
CGAR (%)	**0.77**	**-4.66**

Source: Department of Sericulture

The Karnataka Silk Marketing Board Ltd. promoted by the Government of Karnataka was established in 1979. This was done with the intention of stabilizing the prices of Silk yarn. During 2015-16, the Company has made a turnover of Rs.1145.00 Lakhs as against the turnover Rs.2310.00 Lakhs in the previous year.

2.6 Horticulture Sector

Karnataka is endowed with diverse agro-climatic conditions enabling it to grow a variety of horticultural crops such as fruits, vegetables, flowers, spices and plantation crops. The plantation and horticulture sector plays a vital role in the economy of the state. During the year 2014-15, horticulture crops covered an area of 20.37 lakh ha as against 17.64 lakh ha during 2007-08 (Table 2.13). Though the area under horticultural crops had a share of less than 15 per cent of the net cultivated area of the state, the income generated from the sector accounts for 40 per cent of the total income from the agriculture sector (Anon., 2017). Karnataka is implementing the National Horticulture Mission (NHM) programme from 2005-06. The coverage of NHM increased from 15 districts in 2005-06 to 30 in 2014-15. The total area under horticultural crops has gone up by 14 per cent between 2010-11 and 2014-15 with area under fruit crops expanding by more than 25 per cent during the period.

Table 2.13: Trends in Area under major Horticultural Crops in Karnataka (Lakh hectares)

Crops	Fruits	Coconut	Spices	Vegetables	Cashew	Flowers	Others	Total
2007-08	2.99	4.6	2.46	4.27	0.70	0.25	2.37	17.64
2008-09	3.18	4.71	2.43	4.21	0.70	0.27	2.50	18.00
2009-10	3.60	4.87	2.66	4.37	0.70	0.27	2.52	18.99
2010-11	3.79	4.97	2.75	4.58	0.73	0.28	2.57	19.67
2011-12	3.69	5.10	2.29	4.20	0.70	0.29	2.58	18.85
2012-13	3.77	5.15	1.71	4.03	0.72	0.30	2.88	18.56
2013-14	3.87	5.10	2.00	4.46	0.71	0.30	2.78	19.22
2014-15	3.94	5.23	2.24	5.03	0.75	0.31	2.87	20.37
CGAR (%)	**3.67**	**1.79**	**-3.67**	**1.32**	**0.71**	**2.88**	**2.75**	**1.47**

Source: GOK, Economic Survey, 2016-17

Horticulture sector has emerged as an important component of the economy of Karnataka state and has contributed more than one third share to the economy of agriculture and allied sectors. In many dry regions of the State, Horticultural crops have evolved as an alternative crop to agricultural crops. Horticulture crops cover an area of 20.37 lakh hectares and the annual production is 173.60 lakh Metric tonnes. The average productivity of horticultural crops in the state is 8.52 Metric tonnes per hectare. The annual value of horticultural products produced in the state is Rs.40002.00 crores and constitutes 35.94% of the total income from entire agriculture sector. The share of Horticultural produces in total GSDP of the state is 4.41%.

Horticulture sector includes production of fruits, vegetables, medicinal plants, plantation crops, flowers, and spices.

2.6.1 Production and Productivity

The production of horticulture crops has been expanding steadily in the state. Driven by factors such as the changing consumption pattern among the population in favour of fruits and vegetables and overseas requirements from the demand side, and the shrinking per farmer availability of cultivable land leading to a shift from extensive cultivation of crops such as cereals and pulses towards to more rewarding intensive fruit and vegetable cultivation on the supply side, and the stimulus received through growing public and investment in horticultural research, development and post-harvest management, the sector witnessed significantly rapid growth since 12th Five Years Plan in the cultivated area, production levels as well as productivity.

Area under various horticultural crops during 2007-08 to 2014-15 is given in Fig 2.4. The area under horticulture crops raise from 17.64 lakh hectares during 2007-08 to 20.37 lakh hectares by 2014-15. Area increased at an average annual growth rate of 1.47 per cent during the period 2007-08 to 2014-15.

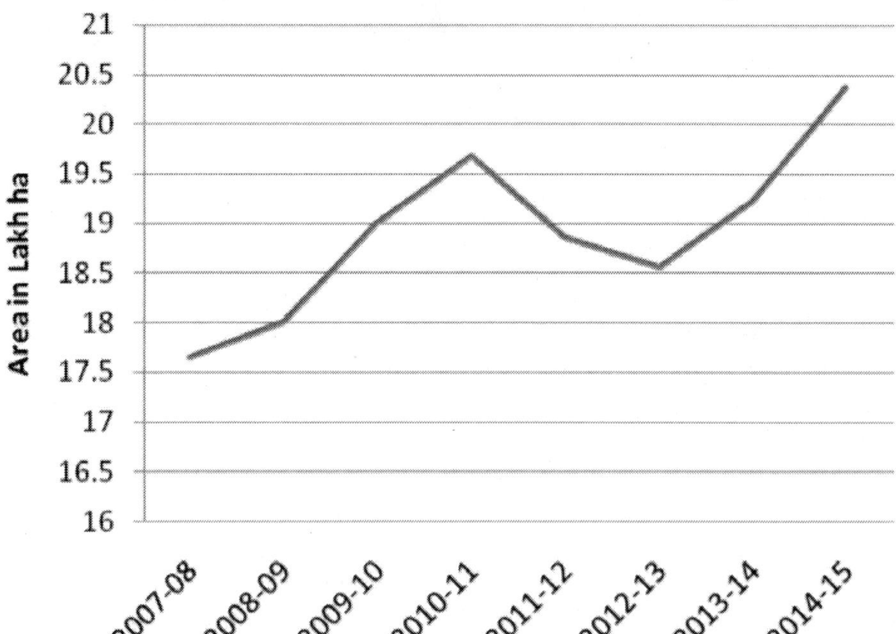

Fig. 2.4: Trends in area under horticultural crops

In order to increase the area of horticulture crops, the department of horticulture Govt. of Karnataka taken up different schemes. Among those, important ones are National Horticulture Mission and National Mission for Oil seeds and Oil Palm. The details of these schemes and progress in implementation are discussed below:

a) National Horticulture Mission (NHM)

For overall development of horticulture, the Government of India started implementation of the National Horticulture Mission (NHM) Scheme in Karnataka since 2005-06. The scheme is being implemented in all 30 districts of the state. At present, this scheme is a sub-scheme of Mission for Integrated Development of Horticulture (MIDH).

Inculcation of protected cultivation method, rejuvenation of the unproductive orchards, rain water harvesting, organic farming, induction of Integrated Nutrition Management (INM) and Integrated Pest Management (IPM), providing post-harvest support, improving marketing, enhancing the technical knowledge and skills of farmers, extension officials and entrepreneurs through an array of training programmes. The main emphases were given on the following programmes in 2016-17 *viz.*, Post-Harvest Management, Protected cultivation, Plantation Infrastructure, Mechanization, Marketing Infrastructure Programmes, etc.

b) National Mission on Oilseeds and Oil Palm (NMOOP)

In India, every year about 144 lakh tonnes of edible oil is being imported at a total cost of Rs.65,000.00 crores per annum. Therefore, in order to decrease National foreign exchange and achieve self-sufficiency in edible oil production, both Central and State Govt. are encouraging oil palm cultivation since 1990-91. The financial sharing pattern between Central and State Governments is in the ratio of 60:40. It is known from survey that in Karnataka, an area of about 2.60 lakh ha has potential for oil palm cultivation. Since, Oil Palm is a water loving crop, its cultivation is being taken up by farmers in the districts which come under Command areas of Cauvery, Bhadra, Thungabhadra, Krishna, Malaprabha and Ghattaprabha through Government approved oil palm entrepreneurs.

For integrated oil palm development, the responsibility of nursery raising, area expansion, distribution of inputs, procurement of Fresh Fruit Bunches (FFBs) and establishment of processing mill has been entrusted to Private partners. Hence, this programme is being implemented in a Public Private Participation (PPP) module. Currently, 5 entrepreneurs are operating in Karnataka. In order to protect the interest of oil palm farmers, oil palm entrepreneurs and oil palm industry as a whole, the State Government has passed the "Oil Palm Cultivation, Production and Processing Regulation Bill" during the year 2013.Currently in the State, around 8600 farmers are cultivating oil palm in an area of about 10400 ha. The annual production of oil palm FFBs is about 15000 M.T, out of which about 2583 M.T of Crude Palm Oil (CPO) is being extracted.

In order to encourage the farmers to take up oil palm cultivation, the farmers are being provided with various kinds of assistance. To establish oil palm plantations, farmers are given input assistance at the rate of Rs. 16,000/-per hectare in the form of inorganic fertilizers for an initial period of four years or pre-bearing period and planting material subsidy at Rs. 8000/- per hectare. Subsidies are also extended for the purchase of various implements, to adopt drip irrigation, inter-cropping, borewell, water harvesting structures and diesel pumpset. Farmers are also taken on exposure visits and imparted training on advance cultivation practices. As on December, 2016, 1,13,685 seedlings have been distributed to the farmers for area expansion.

2.7. Veterinary and Animal Husbandry Sector

The Animal Husbandry and Livestock sector has been playing a significant role in rural economy of the State. In Karnataka, majority of the population is dependent on Agriculture and allied activities. As per the 19th Livestock Census, Karnataka has 2.9 crores of livestock and 5.3 crores of poultry population in the state. During 2015-16, the share of Animal Husbandry in Gross State Domestic Product (GSDP) of Agriculture and Allied activities was 18.78%. As per 19th Livestock Census, the share of Karnataka in all India livestock and poultry population was 5.41% and 7.33%, respectively. The density of livestock in the state was estimated at 151.21 per sq.km and 47468 per lakh human population. The livestock and poultry statistics is given in Table 2.15.

Animal husbandry and dairy farming are important sources of supplementary income to farmers. Livestock also helps in reducing the variability in income of farm families and insulates them from risk. Animal husbandry contributes about 3 per cent to the GSDP of

Karnataka and roughly 22 per cent to the state GSDP generated from agriculture as a whole. Karnataka is a leading state in milk production in the country and has a good network of milk co-operative societies, milk-chilling plants, veterinary hospitals, AI centres, *etc.* Karnataka ranks third in sheep population and sixth in egg production (GoK, 2012). The bovine (cattle and buffaloes) population of the state was 152 lakh in 1997, but declined to 135 lakh in 2003. However, as per the 18th Livestock Census, bovine population has increased to 148 lakh in 2007 (Table 2.14). The sheep and goat population in the state reached 157 lakh in 2007 from 117 lakh during the 17th Livestock Census of 2003, a 33 per cent growth during the five-year period. Poultry population in the state increased by more than 70 per cent between 17th and 18th Livestock Census.

Table 2.14: Trends in Livestock and Poultry Production in Karnataka

(Nos. in Lakh)

S. No.	Species	2003	2007	2012	2003 to 2007		2007 to 2012	
					Change (No.s)	Change Percentage	Change (No.s)	Change Percent-age
1	Cattle							
	Crossbred	1612570	2177425	2275400	564855 (+)	35.02	97975 (+)	4.31
	Local	7983360	7412191	7241084	571169 (-)	7.15	171107 (-)	2.31
	Total cattle	9595930	9589616	9516484	6314 (-)	-----	73132(-)	-----
	Buffaloes	4023832	4072233	3470505	48401 (+)	1.2	601728 (-)	17.34
	Sheep	7272241	8950584	9583761	1678343 (+)	23.07	633177 (+)	6.61
	Goats	4491473	5387329	4796147	895856 (+)	19.95	591182 (-)	12.33
	Pigs	320433	241212	304798	79221 (-)	24.72	63586 (+)	20.9
	Others	39038	34672	35711	4366 (-)	11.2	1039 (+)	2.9
	Total livestock	**25742947**	**28275646**	**37223890**				

Source: Livestock Census Reports, 2003, 2007 and 2012 Govt. of Karnataka

There were 10,534 active dairy co-operative societies with 20.40 lakh active members as on September 2015. The State operates 23 dairy processing plants with a total capacity of 32.25 lakh litres per day and has 42 milk-chilling centres with a capacity of 15.15 lakh litres per day. Karnataka also has 6 product dairies with a capacity to produce 92 MT of milk powder per day. As on March 2015, the state has 4110 veterinary institutions among them 371 Vet. Hospitals, 1942 Vet. Dispensary, 1180 Primary Vet. Centre, 228 A I Centre, 174, Mobile Vet. Centres and 215 others (Anon., 2017).

2.7.1 Livestock and Poultry Production

In milk production, India ranks second in the world and during 2015-16, Karnataka State ranks eleventh among Indian States. The production of milk in the State was 6.34 million MT during the year 2015-16. Details of the production of the major livestock products namely, milk, meat, wool and eggs are shown in Table 2.15 and Table 2.16.

Table 2.15: Trends in Livestock and Poultry Statistics in Karnataka

(No. in crore)

Particulars	Livestock Census			
	2003	2007	2012	Percentage change in 2012 over 2007
Cattle	0.95	1.05	0.95	-9.52
Buffalo	0.40	0.43	0.35	-18.60
Goat & Sheep	1.18	1.57	1.44	-8.28
Other Livestock	0.30	0.24	0.16	-33.33
Total Livestock	2.83	3.29	2.90	-11.85
Poultry	2.44	4.24	5.34	25.94

Source: Department of Animal Husbandry Annual Report, 2016-17.

Table 2.16: Trends in Livestock and Poultry Production in Karnataka

Item	Unit	2013-14	2014-15	2015-16	2016-17	Percentage change in 2015-16 over 2014-15
Milk	'000 Tonnes	5997	6123	6344	5490	3.61
Meat	Tonnes	169894	181312	196600	179932	8.43
Wool	Tonnes	7755	8821	8191	5681	-7.14
Eggs	No. in crores	412	439	476	418	8.66

Source: Department of Animal Husbandry, GOK, Annual Report, 2016-17.

2.8 Fisheries Sector

Karnataka has a 300-km coastline with a continental shelf of 27,000 sq km and 5.60 lakh hectares of inland waters. This provides immense scope for the development of fisheries. In addition to this, about 8,000 ha of brackish water area provides good scope for shrimp farming. The total fish production in Karnataka was around 2 lakh tonnes in the early eighties and reached more than 5 lakh tonnes in 2014-15, of which 60 per cent was contributed by the marine sector and 40 per cent by the inland fisheries sector (GoK, 2014-15). The growth in marine and inland fish production is presented in Table 2.17.

Table 2.17: Trends in Fish Production in Karnataka

Year	Marine Fish Production		Inland Fish Production		Total Production	
	Quantity (Lakh tonnes)	Value (Rs. in lakh)	Quantity (Lakh tonnes)	Value (Rs. in lakh)	Quantity (Lakh tonnes)	Value (Rs. in lakh)
2004-05	1.72	45,873	0.80	21,582	2.52	**67,455**
2005-06	1.76	46,598	1.21	32,321	2.97	**78,919**
2006-07	1.68	55,143	1.24	37,176	2.92	**92,319**
2007-08	1.75	51,787	1.22	36,637	2.97	**88,424**
2008-09	2.18	82,024	1.44	57,487	3.62	**1,39,511**
2009-10	2.49	90198	1.59	63730	4.08	**153928**
2010-11	3.41	133565	1.86	77348	5.27	**210913**
2011-12	3.47	150630	1.99	80016	5.46	**230645**
2012-13	3.57	201668	1.68	84605	5.26	**286273**
2013-14	3.57	130010	1.98	79361	5.26	**209371**
2014-15	3.89	140911	2.24	89783	5.55	**230694**
CAGR (%)	**7.5**	**15.4**	**11.92**	**22.58**	**9.01**	**17.99**

Source: www.indiastat.com

In Karnataka state, fisheries has attained as an important sector contributing to fish production besides source of foreign exchange earner. Fish being one of the sources of quality protein, use of fish as food helps in eradiation of malnutrition among rural population. Fisheries sector contributes to production of quality animal protein required for human being besides provides more employment opportunities. The vast marine, brackish water and inland freshwater resources are the source of fish production in the State Karnataka

State has 320 km. long coast line along with 27000 Sq. km continental shelf area, 5.65 lakh hectares of various inland water resources and has vast scope for fisheries development. The brackish water area of 8000 hectares also provides good scope for shrimp/fish culture. There are about 9.61 lakh fishermen in the state, of which 3.28 lakh fishermen in marine and 6.33 lakh fishermen are in inland who are involved in various fisheries activities. During the year 2015-16, the total fish production of the state is 5.80 lakh tones. Karnataka is in 6th position in marine fish production and 8th position in inland fish production, when compared to fish production in the country. The total fish production during 2016-17, upto end of December 2016 is 3.93 lakh tonnes.

2.8.1 Fish Production

The annual fish production in Karnataka has shown a considerable increase from 2.51 lakh metric tonnes in 2004-05 to 5.80 lakh metric tonnes in 2015-16. Details of fish production for the last 4 years are given in Table 2.18 and Table 2.19.

Table 2.18: Details of Fish Production in Karnataka (MTs)

Year	Marine	Inland	Total
2013-14	35 7358	197952	555310
2014-15	38 9822	223419	613241
2015-16	41 1762	168828	580590
2016-17	26 2751	130000	362751

Source: Department of Fisheries, GOK, Annual Report, 2016-17.

Table 2.19: Inland fisheries resources in Karnataka

Type	Area/Length
Department Tanks	1.77 lakh ha
Gram a Panchayat Tanks	1.21 lakh ha
Reservoir	2.72 lakhh a
Rivers	5813 km
Canals	3187 km
Brackish Water Areas	0.08 lakh ha
Production Potential	4.02 lakh MT

Source: Department of Fisheries, GOK, Annual Report, 2016-17.

❑❑❑

Chapter 3

Methodology

The present study is conducted to address the developmental needs of the economy with special reference to education in agriculture and its allied sectors, which have an important place in the Karnataka state economy due to their role and potential in providing employment to skilled, semi-skilled as well as unskilled persons. Skill needs are changing and hence, a need for analyzing the skill gaps, the institutional mechanism in imparting education and assess the future requirements of the sector in the context of human resources.

The objectives of the study are:

- To estimate the demand for services of agricultural graduates in public and private sector
- To assess the employment opportunities for agricultural graduates turning out in the state.
- To review ICAR and SAU schemes for HRD of students in SAU for improvising
- To identify gaps in quantity and quality of supply of skilled man power in agriculture
- To suggest measures and strategies for the development of skilled human resources in agriculture and allied sectors.

This study covers agriculture and allied sectors education in Karnataka (Fig. 3.1). In order to address the above objectives, the following methodological approach was adopted to make an assessment of the future human resources needs in agriculture. This chapter:

- Provides a conceptual framework for human resource assessment
- Review of available approaches and methods of human resources forecasting
- The approach and methodology adopted for the present study

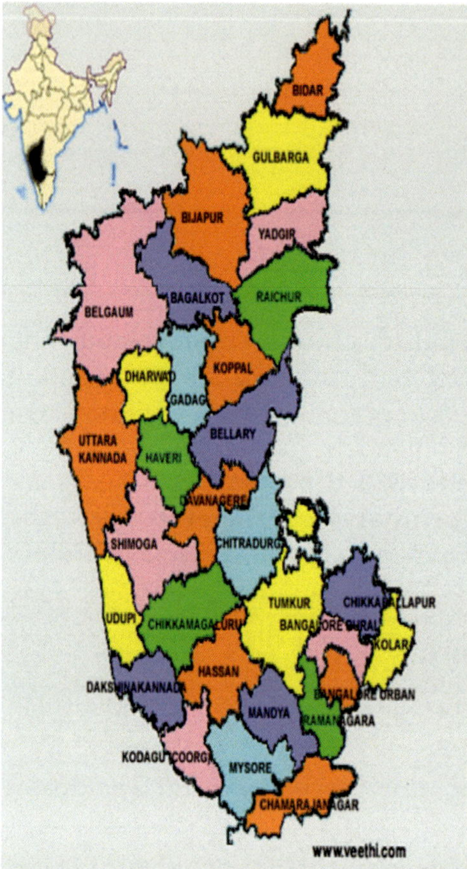

Sl. No.	Name of the Place	No. of Universities	No. of Colleges
1.	Bagalkot	1	1
2.	Belagavi, Arabhavi		1
3.	Bengaluru	1	4
4.	Bidar	1	2
5.	Bijapur		1
6.	Chikkamagaluru, Mudigere		1
7.	Chitradurga, Hiriyur		1
8.	Dharwad	1	2
9.	Hassan		1
10.	Haveri, Hanumanamatti		1
11.	Kalaburgi		2
12.	Kolar, Chintamani		1
13.	Koppal, Munirabad		1
14.	Kurg, Ponnampet		1
15.	Mandya		1
16.	Mangalore		1
17.	Mysore		1
18.	Raichur	1	2
19.	Shimoga	1	2
20.	Uttar Kannada, Sirsi		2
21.	Yadagir, Bheemanrayagudi		1

Fig. 3.1: District-wise distribution farm universities and colleges in Karnataka

3.1 Conceptual Framework

The broad conceptual framework of the study, comprising distinct phases, with coverage of agriculture and allied sectors in the state are schematically presented in Fig. 3.2. The study aimed at assessment of human resource requirements in agriculture and allied sectors, which necessitated state-wide surveys and consultations with various stakeholders.

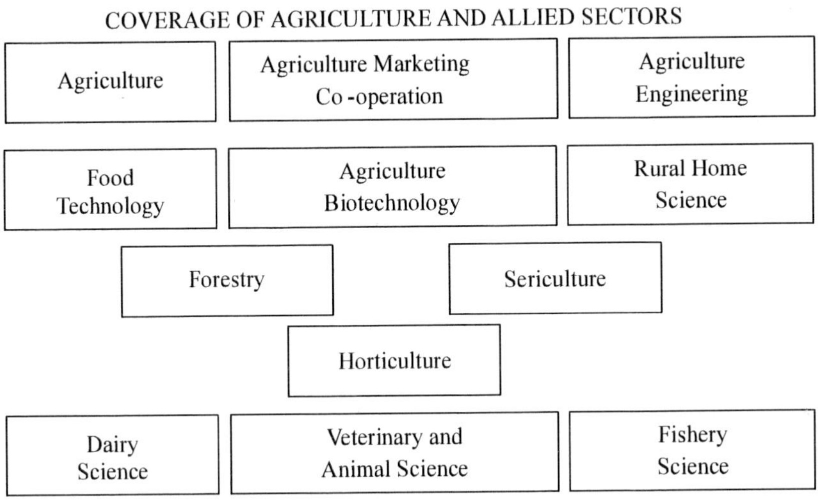

Fig. 3.2: Schematic diagram of the areas agriculture and allied sectors covered in the present study

3.1.1 Human Resources Supply-Demand Process

Both human resources demand and supply can be viewed in terms of both flows as well as stocks. When considered in terms of flows, these terms imply net additional demand and supply during a time period, say a year. In terms of stocks, they imply the total quantum of human resources deployed & the total stock of economically active human resources available at a particular point in time. A graphical representation incorporating the essentials of the human resources demand and supply processes is given in the Fig. 3.3, which presents the manner in which all

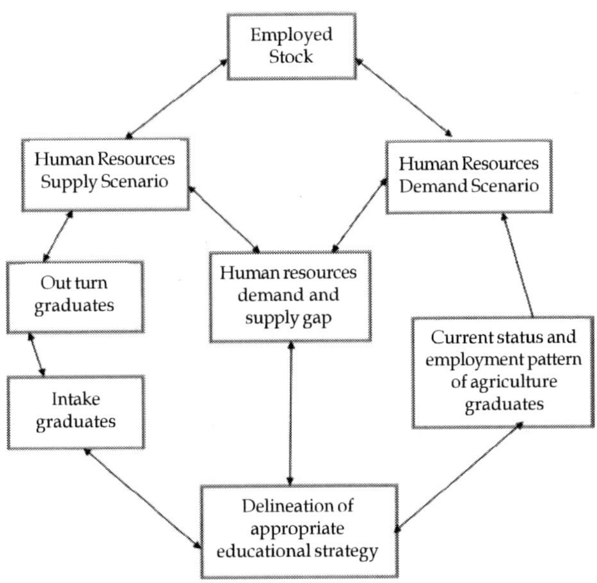

Fig. 3.3: Supply-Demand Process

the data collected, would lead to integrated framework for analysis of human resources demand and supply gaps.

3.2 Different Tools /Approaches Available for the Forecasting

There are various studies are being conducted at national and international levels. The brief reviews of past studies are present the following sections:

3.2.1 Qualitative Forecasts

Qualitative changes likely to take place in future in the human resources, demand and supply are captured primarily through qualitative data. The qualitative methods do not aim at exact quantification of future human resources demand in terms of numbers, but ascertain aspects such as gaps between demand and supply in skill content, newly emerging skill needs, structural factors affecting the demand-supply balance, direction of future changes, *etc.* Such methods include collection of desired qualitative information through structured or unstructured questionnaires in an employer's survey, or eliciting the opinions of experts in the field through key informants approach, focus group discussions covering various quality (as well as in broad terms, certain quantity) aspects, Delphi technique, etc.

3.2.2 Quantitative Forecasts

Over the past half century, a variety of human resources forecasting techniques have been employed in different states, countries and in different situations of data availability. These include:

- Area and service based approach
- Employer survey methods
- Norm-based methods
- Time-series and regression (linear/non-linear, simple/multiple) models
- Other econometric models
- Mathematical models
- Rate of returns approach
- Parnes's manpower requirements approach

- System dynamics (SD)
- Causal loop model (CLM)

3.2.2.1 Area and Service based Approach

In this approach requirement of man power can be assessed based upon the standard references given by FAO, such as number of skilled persons required per unit area or service required per unit population. There by it is possible to predict or forecast the future need of human resources in a given country or state. However, this method has limitations since extension could also be done through various modern methods, for instance use of ICT in agriculture.

3.2.2.2 Employers' Surveys of Requirements

A straight-forward method of ascertaining future human resources requirements is to approach the employing agencies directly, either on census or sample basis, and asking for information on their present human resources deployment, skill shortages and the anticipated needs over the forecast period. The method has the advantage of getting information, based on the experience and intimate knowledge of the employers about the future labour markets and anticipated technological changes in the industry. However, this approach is hinder by the fact that few employers would be in a position to foresee their human resources requirements beyond a short period, and even if they are able to, would not always be willing to part with the information.

3.2.2.3 Norm-based Forecasts

In the case of development programmes with standard manning patterns established in relation to the tasks to be performed, future human resources needs can be worked out by applying the staffing norms to the anticipated or planned expansion of the programmes. This approach is applicable mainly in the government sponsored programmes like Community Development, Intensive Child Development Scheme, etc., but has limited or no use in the case of private sector activities. Norm-based forecasts are also useful in arriving at requirements of categories such as teachers (based on projected number of students and anticipated pupil-teacher ratios) or doctors (based on projected population and desirable density of doctors per thousand of population).

3.2.2.4 Time-series and Regression Models

Time-series models enable making human resources (or for that matter of any economic variable) forecasts on the basis of a study of the historical pattern of changes (usually trends) in the past. Unlike these models, which make use of past data on human resources only, regression models establish relationships between human resources and other safely predictable variables, which are associated with human resources variables. Both these types of models, however, require long (spatial or temporal) series of data, not usually available in the field of human resources planning in developing countries. For example, for applying a regression model to forecast requirements of agricultural human resources, one would require a sufficiently long time series on the size of agricultural human resources on the one hand and on the levels of variables that influence the requirement of such human resources like technology, extension network, etc.

3.2.2.5 Econometric Models

More complicated econometric models postulating the interplay of a number of variables through a set of structural equations have been developed for forecasting human resources. For example, the Timbergen-Jos model comprises six linear equations relating volume of production, number of students at two levels of education, and the members of the labour force and those entering the labour force during the past six years. The BACHUE models developed by ILO during the 1970s include three sub-models, the demographic, the economic and the income distribution sub-models. The first sub-model tacks the population by age, sex, location and education, the second generates final demand and value added by sector using input-output table, and the third predicts employment and distribution of income. Applicability of these models is again restricted by data availability.

3.2.2.6 Mathematical Models

These models include Markov models, Simulation models, and System Dynamic Models. Markov chain models are extensively used, and are used on the concept of transition probability matrix and usually formulated to estimate long-term study state equilibrium conditions. In general, these exercises pertain to two distinguished dimensions: human resources supply and human resources demand. Various methods and tools are developed over the time for forecasting.

Simulation models imitate and assess the impact of various management scenarios, and answer questions, regarding human resource management that determine actual requirements primarily at unit level. System Dynamics (SD) modelling methods developed by Jay W. Forrester is a powerful methodology to understand the systems, their variables and parameters, inter-linkages; and to develop a model using both qualitative and quantitative variables (Mohapatra *et al.*, 1994). Vizayakumar (1990) has explicitly stated the important features of system dynamics as the ability to (i) dynamically model complex relationships of a large number of variables, (ii) consider many related aspects of a problem resulting in a holistic approach, (iii) explicitly model qualitative factors, (iv) experiment with policy alternatives, (v) generate future scenarios, (vi) incorporate features of both the explanatory and normative approaches, (vii) model a wide variety of real world situations. This method was effectively used for human resources assessment in CSIR (Mandal *et al.*, 1992, and Shivanagaraju, 1998 & Mohapatra *et al.*, 1990), for food production (Mandal, 1993), and for oilseed system (Rama Rao *et al.*, 2000) and for soybean system (Bisht *et al.*, 1996).

3.2.2.7 Rate of Returns Approach

In this approach, the rates of return to investments in different streams of education are computed by assessing the lifetime earnings foregone. Investment decisions are made on the basis of the relative rates of return from different streams. This approach helps in identifying as to which stream of education results in a better rate of return on investments. It does not help, however, in arriving at a quantitative assessment of human resources requirements.

3.2.2.8 Manpower Requirements Approach (Parnes' Model)

A general approach for forecasting human resources requirements for the purpose of educational planning was originally propounded by Parnes (1962) in the context of Mediterranean Regional Project during the early 1960's.

3.2.2.9 System Dynamics (SD)

A methodology of system enquiry (Wolstenholme and Coyle, 1983) is a theory of structure and behaviour of systems that helps in analyzing and representing the interactions governing the dynamic behaviour of complex socio-economic systems (Forrester, 1961). It can handle complex feedbacks and delays present in the system in predicting the system's behaviour over time. To develop a system dynamics model, a causal loop model (or CLD) is developed initially. Causal loop diagrams depict the causal relations that exist between the variables in a system through the use of text, arrows and symbols (Stepp *et al.,* 2009). A causal relationship between two variables is positive, if they move in the same direction and negative, if they vary in the opposite direction. A causal loop is reinforcing, if it has zero or an even number of negative causal relations and which result in reinforcing the behaviour of the system. A causal loop is balancing, if it has odd number of negative causal relations and which stabilizes the system behaviour over time. System dynamics aids in analysing the effects of alternate policies on the system's behaviour before implementing them.

System dynamics have been used in manpower planning and forecasting studies all over the globe. Suitability of SD in manpower forecasting is cited by Jantsch (1972; 1973), Khoong (1996), and Martino (1980). Parker and Caine (1996) bring out the advantages of utilising system dynamics methodology over sophisticated mathematical modelling techniques in human resource forecasting. System dynamics was used for manpower planning and forecasting in steel plants (Roy and Koul, 2009), health services (Chung *et al.,* 2010), enterprises (Wu *et al.,* 2003), and information security industry (Park *et al.,* 2008). Sterman (2002) suggested that complex systems require a mastery of concepts such as stocks and flows, feedback, delays, and non-linearity.

3.2.2.10 Causal Loop Model (CLM)

It is a dynamic integrated demand-supply model for manpower forecasting in agriculture sector in India. The model is built on a system dynamics framework. Initially, a causal loop model (or casual loop diagram) is presented to capture the structural relationships existing between manpower supply and demand and its determinants (Figure 3.4). Details of causal loop model on feedback loops, loop polarity, loop name and loop components are depicted in Table 3.1.

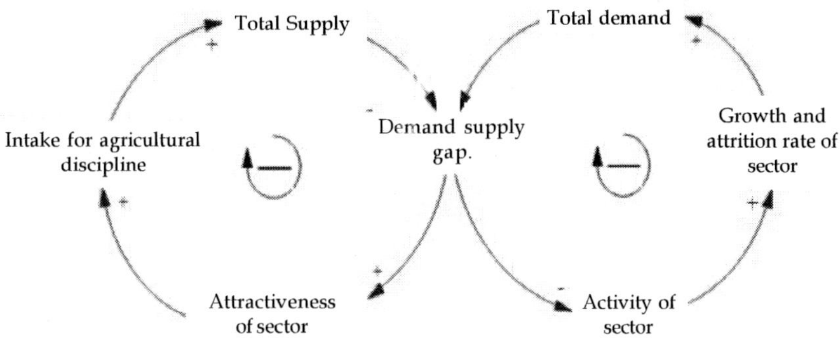

Fig. 3.1: Causal loop diagram for the manpower forecasting model

Table 3.1: Details of causal model on feedback loops, loop polarity, loop name and loop components

Loop ID	Balancing (-) or reinforcing (+) loop	Loop name	Loop components
B1	Balancing	Supply enhancement loop	Total supply, Demand-supply gap, Attractiveness of sector, Intake for technical courses
B2	Balancing	Demand fulfillment loop	Total demand, Demand-supply gap, Activity of sector, Growth and attrition rate of agriculture sector

The causal loop diagram in Figure 3.1 shows two negative feedback loops–one related to supply enhancement (loop B1) and the other related to demand fulfillment (loop B2). The ultimate goal is to create a situation where the demand and supply equilibrates. However, dynamicity of the loops creates imbalances all the time.

Agriculture manpower in India comes mainly from educational institutions (undergraduate and graduate colleges). The supply constitutes the graduating students from these colleges and universities in three categories - degree graduates, post-graduates and doctorates. The demand is generated from the attrition of currently employed stock and additional requirements due to the growth of the agriculture sector employing the graduates. Demand-supply gap affects the attractiveness of the sector, which subsequently affects the intake in the technical institutes in the country and also negatively affects the central activity of the sector.

3.3. Review of Past Studies in India for the Forecasting of Human Resources

The manpower planning process includes forecasting the future demand and supply of manpower and then developing action plans to reconcile the discrepancies between demand and supply. Reliable manpower demand and supply forecasts can provide a basis for making better decisions for avoiding redundant investments, achieving efficient and balanced growth of an industry, and in developing an economy. Much of the literature on demand analysis is devoted to manpower forecasting at the micro level, while manpower forecasting at the macro level is equally important, when it comes to the economic development of a country and state (Balagopal *et al.,* 2010). Ever since the planning era began in the country, India too has been attempting forecasts of requirements of various human resources categories as a part of its developmental planning. Some important efforts in this direction specifically in relation to agricultural human resources are summarised below.

3.3.1 Forecasts in 1970s

Human resource requirement assessment during earlier mainly done by area and service based approaches, that is on the basis of certain assumed staffing norms for organizations at block and district levels (e.g. one village level worker for every 10 villages, 3 to 4 agricultural graduates for every block, and 4 to 5 subject-matter specialists in agriculture in every district) FAO, (1970). Replacement needs were calculated at 3 per cent of the employees in position on the assumption of an average working life of 30 to 35 years and even distribution of the stock of personnel across all age groups. Demand estimates were similarly arrived at for various allied sectors like animal husbandry, dairying, fisheries, *etc.*

3.3.2 Forecasts in 1980s

The two studies conducted in 1984 by Institute of Applied Manpower Research (IAMR). These studies were some limitation because 1) they covered only the requirements of human resources directly involved in agricultural planning and projects, 2) only four government departments, *viz.,* Agriculture, Animal Husbandry, Horticulture and Sugarcane were covered and private sector was excluded. The estimation was based on data collected from the relevant government departments and agricultural personnel through a set of questionnaires. Another study by the Ministry of Food and Agriculture, GOI, in this approach earlier studies limitation was overcome that is the requirements of the private sector were explicitly covered through a survey of agro-industries.

The Institute of Applied Human resources Research carried out in-depth study during 1986-87, by collecting available information and Annual Reports from Agricultural Universities in the country to estimate the requirement of graduates and post-graduates in Agriculture and Allied sciences over a period of ten years.

3.3.3 Forecasts in 1990s

During 1990s, studies on Human Resource Development were carried out in Andhra Pradesh, Tamil Nadu and Haryana. Tata Consultancy Services, Hyderabad, conducted study, in Andhra Pradesh and Tamil Nadu to study the employment trends of the output of the State Agricultural University, forecast agricultural human resources requirements in Andhra Pradesh upto the year 2010 in both technical and general categories, to study the technical competence of the University alumni to meet the demand in relation to curriculum and education system, and to provide a human resources model to present the database on human resources supply and demand projections. They suggested that a regular human resources information system could be updated for further improvement of the education system.

M/s A.F. Ferguson & Co. conducted study for Haryana to identify the present status and emerging human resources requirements of agriculture, horticulture and allied sectors in farming, agro-industrial and services sectors, demand scenario for agri-professionals at state, regional and national levels, self-employment opportunities, required modifications in curricula for better absorption, *etc.* They concluded that about half of the agricultural graduates are currently employed in government and one- fifth in corporate sector. The share of government sector is declining as future opportunities are limited to retirements only. However, the prospects of employment in the corporate sector are bright, about 90 per cent of the agricultural graduates are employed in agriculture related activities and only 10 per cent are in un-related activities or have gone abroad, employment of agricultural graduates in financial institutions has reached a stagnation point. Majority of fresh employment opportunities in rural credit disbursal process go to plain graduates or MBAs with specialization in finance, less than 2 per cent of the agricultural graduates are presently self-employed and this sector's potential needs to be evaluated.

3.3.4 Forecasts in 2000

IAMR conducted national level study titled "National Human Resources Needs Assessment in Agriculture and Allied Sectors" in 2001. The study collected data from 149 employers and 3,661 individual employees, 75 universities and colleges, 1,747 agricultural graduates passed out, and 93 experts in the field. The study recommended, inter alia, a) demand projections based on input-output framework b) a computerised information system on agricultural human resources at state and national levels, c) annual tracer studies covering graduates three years after graduation, and d) a directory of agricultural and allied scientists.

3.3.5 Forecasts in 2010

During 2010, National Academy of Agriculture Research Management, (NAARM, Hyderabad) collaboration with Institute of Applied Manpower Research (IAMR, New Delhi), conducted the study entitled Assessment of Future Human Capital Requirements in Agriculture and Allied Sectors, to assess the trend in supply-demand of trained human resources in agriculture, and to develop a system dynamics model for forecasting and evolve prospective human resource development strategies. Outcome of the study shown that the shares of various segments by employment are: 33 per cent in government, 44 per cent in private, 10 per cent in financial, 4 per cent in research and academic and 9 per cent in others. The major shift in the past three decades is declined in the share of public sector in employment, which may be due to freezing employment in government sector as well as expansion of opportunities in the private sector. This is in tune with the emergence of commercialisation as well as diversification. The gaps between demand and supply of human resources in agricultural and allied sciences even to the tune of 50 per cent or more. This is true across the board, though the shortfall is high in the case of rapidly growing horticulture, dairy and fisheries sectors, and less serious in others. Dairy, fisheries, veterinary and horticulture are the future engines of growth, which would have a bearing on trained human resources requirements. At present, the existing education system is producing about 24,000 graduates per year with crop sciences contributing 2/3rd of it. The projections indicate that by 2020 the annual out turn required would have to be about 54,000, i.e. the demand supply gap would be 30,000. Discipline-wise, the additional annual requirements of out turns are expected to be: Agriculture 9335; Horticulture 7153; Forestry 1116; Dairy 3005; Veterinary & Animal Husbandry 4989; Fishery 2181; Agriculture Engineering 1749; and Agri-Biotechnology 305. During the last decade, out turn of graduates and above grew at the rate of 5 per cent and in the coming decade, it is projected to double the growth rate to meet the demand. An important aspect of educational planning is to ensure availability of faculty in adequate numbers and quality with institutional arrangements for systematic upgradation of their knowledge and skills. About 40 per cent of the faculty positions in the agricultural universities have remained vacant for long periods of time raising questions about the quality of education.

There is a strong demand from all the stakeholders for skill-specific education with clarity of basics as well as hands-on technical expertise. In other words, there is a need to develop functional skills among the students in educational institutions almost in all the sectors of agriculture.

Assessment of Technical Manpower Requirements in Agriculture Sector in India (Balagopal *et al.*, 2010). They suggested that there is need to 50 per cent increase in the intake capacity of the technical institutions is proposed which is thereafter evaluated with anticipated sectoral growth rates of 3 per cent and 4 per cent, respectively. To alleviate the shortage of technical manpower substantially and make the further interventions in balancing the technical manpower supply and demand in the long run.

Nanda *et al.* (2010) conducted study on human resource development for agricultural sector in India. Human resource for the agricultural sector in different sources of employment, *viz.*, government, private (including corporate), academic, financial institutes, non-governmental organizations, self-employment, and others in India. Besides projecting an overall scenario for continuation of current agricultural education policy and trends, they suggested that current curriculum with 80:20 proportion of technical to soft skills.

The analysis shows that in the coming years, the private sector will emerge as a major employer for the graduates of agriculture and allied sciences. The supply of human resource is projected to be around 16,000 in 2020 against a demand of nearly 15,000. Though the growth in government sector is assumed to be modest at 2.5 per cent, positive and healthy growth in other sectors drives, the model to reduce demand supply gap with time. The annual supply-demand gap, declines from 27.1 per cent in 2001 to a minimum of 5.93 per cent in 2019 and thereafter registers a slow growth to 6.1 per cent in 2020.

Vinay *et al.* (2016) conducted the study in fisheries education in Karnataka: trends, current status, performance and prospects. In India, Fisheries education first took root in the college of fisheries, Mangalore and college of Fisheries, Mangalore and the college is the alma mater of many scientists who have contributed to today's well-advanced science and technology in fisheries field both at national and international levels. Until now, the college has produced 1295, 605 and 132 students of UG, PG and Ph.D., respectively and the estimated CGR for UG shows 3.3 % (2002-2014) positive growth. The forecasting of fisheries graduates shows that there will be an increasing trend in the future. Thus, proper utilization of this skilled manpower can improve fisheries production both for the marine and inland sectors.

3.3.6 Forecast Studies by Others

Apart from the above studies, which cover the entire spectrum of agricultural human capital some sector specific assessments have also been made from time to time. At the skilled worker level, Pandit Sunderlal Sharma Central Institute of Vocational Education (PSSCIVE), Bhopal evolved certain norms to assess the requirements of skilled human resources, particularly at lower levels, in different sub-sectors of agriculture (Sacheti and Salooja, 2000). Studies on human resources requirements in fishery sector were conducted by various researchers primarily using normative methods (CIFE, 2000). Similarly in poultry sub-sector, some attempts were made to forecast human resources demand, mainly on the basis of institutional out turns and the resultant stocks (Sasidhar, 2009).

The above brief review of some of the major studies on the agricultural human resources bring out the fact that forecasting of human resources demand and supply has been an adhoc exercise, and each time, the researcher has to grope for data. Non-availability of good data and in good time, hinders application of sophisticated techniques employed in the US, Canadian or Irish occupational forecasting systems.

3.3.7 Limitations in Recent Studies

There have been two studies in recent years that have attempted national level forecasts of requirements of agricultural human resources. The first one is the IAMR's national study and the other, the study by M/s A.F.Ferguson. The latter study, even though primarily focused on Haryana, nevertheless attempted national forecasts as well. Both were conducted roughly about the same time, around the year 2000, and the forecast horizon was 2010 in the case of IAMR's study and 2009 in the second study. There is, however, one important difference. While the IAMRs projections were, stock projections, those in the other study were flow projections. Thus, while IAMR projected the total number of agricultural graduates and post-graduates required by the market by the year 2010, the Haryana study estimated the annual additional number required in the year 2009. As mentioned above, the two sets of projections are not directly comparable as one represents the stocks required and the other the annual additions required. For purposes of comparison, therefore, the IAMR forecasts for 2010 have been converted first to relate to 2009 by linear interpolation between 2005 and 2010 and then as flow variable by adding the annual expansion demand and replacement demand (estimated at 1.3% of the stock). These are shown in the last row of the table. It will be seen that the trend based forecast of IAMR for 2009 (suggesting an annual outturn of about 10,000 graduates and post-graduates) is fairly comparable with the forecast from the Haryana study. This, however, appears to be much lower than the current annual output from the agricultural educational institutions. It may be mentioned that the trend-based forecast of IAMR is obtained by extrapolating the growth of employed stock between 1981 and 1991, a period very much in the past.

3.4 Review of Past Studies in India for Assessment of Quality of Human Resources

In order to provide a conceptual framework for human capital assessment, pertinent articles were collected and reviewed. The available approaches and methods of human resources forecasting were made, based upon reviews of past studies and justified the gaps therein and outlines the approach and methodology adopted for the present study. The abstracts of selected studies are as follows:

Annon (2005) studied the agricultural human resource development in India and suggested that strengthening and development of agricultural education was attempted through Centres of Advanced Studies, Niche area of excellence, University level-textbook writing, National Fellows and Professional Chairs, infrastructure upgradation, faculty competence building, accreditation, incentives through best teacher award, scholarships to students and their practical training in laboratory and real life field situations. Entering into MoU with the IGNOU for co-operation in agricultural education in distance mode and co-operation with the ISRO, IGNOU and MHRD for utilization of EDUSAT by the ICAR and SAUs for agricultural education are the new initiatives, which have immense potential for enriching agricultural education as well to reach the unreached.

Rama Rao *et al.* (2011) assessed the future human capital requirements in agriculture and allied sectors in India. They reported that, compared to other streams of science and technology, the growth in agricultural education is relatively low (3 per cent) during the period 1970 -2000. In the past decade, the growth has accelerated to 5%, the result of which can be seen from the increased out turn of students in the last five years especially with the emergence of private sector participation in agriculture education. Considering the current developments and projected growth of various sectors of agriculture in the next decade, agricultural education is likely to experience above 5 per cent growth per annum. On this scenario, there will be a gap of 37 per cent in demand-supply.

Rashmi Agrawal *et al.* (2011) reported that at the current rate of outturn is less than 65 per cent of the intake capacity. Though full capacity utilization is not possible, the system efficiency can be improved to produce another 15-20 per cent additional graduates from the existing system. The expansion may have to consider this factor. The demand of professionals at PG and PhD level would be met through increase in number of sanctioned seats in existing colleges. At UG level, considering the current intake capacity and large gap in demand-supply, there is a need to establish about 150 new colleges.

Ashish Makwana (2013) conducted the study on Agricultural Education in India: Challenges & Prospects, and mentioned the importance of agriculture education in India. After independence, state agricultural universities were established in all the states to impart education in the field of agriculture. Indian Council of Agricultural Research, New Delhi regulates and ensures effective dissemination of agricultural education. Agricultural education being a professional education, it is necessary to take remedial measures so as to ensure food sufficiency for our growing population.

Annon (2013) quality human resource development is crucial for addressing the challenges in sustaining agricultural productivity in diversified agro-ecological regions. To address the challenges of agricultural growth and enhancing the benchmark quality standards of higher agricultural education across the State Agricultural Universities (55 SAUs), Deemed-to-be Universities (5 DUs), Central Universities (4 CUs with agricultural faculties), and Central Agricultural University (1) the Education Division of ICAR is playing a pivotal role by providing logistics and catalytic financial support for infrastructure development including domains of governance, development of uniform course curricula and capacity building. This has been possible through the XI Plan scheme of "Development and Strengthening of Higher Agricultural Education in India" through various sub-schemes and programmes.

Sandra Damico (2010) reported that Cambodia's higher education system has grown rapidly over the past 5 years, in terms of number of Higher Education Institutions (HEIs) and student enrolments with high concentration in the capital city and main provinces. Practical solutions to improve education quality can be done in the following ways: 1) Strengthening accreditation (of HEIs and course curricula) within a strategic framework that addresses a workforce vision and how higher education will contribute to this development; 2) Before issuing foundation year course certificate, Accreditation committee of Cambodia should impose stricter measures on accreditation and quality assurance process by carefully evaluating six assessment criteria: i). Department of foundation year course; ii) Strategic plan for foundation year course; iii) Foundation year course curriculum; iv) Teaching staff in foundation year course program; v) Physical facilities and learning resources; vi) student admission; 3) Improve teacher standards through providing support to universities and building links with private sector to that have working experience and practical implementation of theories learnt during university. Building a qualification framework for teachers and regular updating of skills and methodologies through a centralized coordination mechanism - either a university association or the DHE itself.

Ayyappan *et al.* (2007) made an interface in agricultural transformation through public-private partnership. All partnership efforts for agricultural transformation should ultimately lead to economic prosperity of the farmer. This means higher productivity of the crops with newer technologies, be it superior hybrids; exploiting phenomenon of hybrid vigour or transgenic crops, where an economic trait is incorporated from an unrelated species. Innovation is one thing and reaching of the technology to the farmer is another. Both are very important. All along in the seed industry concerns have been on Food Security and Nutrition Security for the nation. To achieve these goals, partnerships at various levels with different objectives have to be developed amongst various stakeholders, which shall bring prosperity to the nation through the prosperity of the main stakeholder, the farmer. For the success of the partnership with Public Institutions and Private Organizations, identifying the strengths of the partner and the areas of complementarities are very important.

Nanda *et al.* (2004) system dynamics model was developed for dynamic analysis of human resource for the agricultural sector in different sources of employment, *viz.*, government, private (including corporate), academic, financial institutes, non-governmental organizations, self-employment, and others in India. Besides projecting an overall scenario for continuation of current agricultural education policy and trends, the paper analyses simulated results from the model for the current curriculum with 80:20 proportion of technical to soft skills. The analysis shows that in the coming years, the private sector will emerge as a major employer for the graduates of agriculture and allied sciences.

Swaminathan (2005), agricultural education in India is faced with one of the biggest challenges. It has to identify its role in equipping the human resources for enhanced agricultural productivity and sustainable use of natural resources. Agricultural colleges and universities were initially assigned to disseminating scientific knowledge and skills to the farming community and to train them to use such skills for better output. As a backup for such a mission, agricultural research was encouraged to adapt the scientific knowledge to suit to the realities of rural societies. However, these initiatives could not keep pace with the fast changing scientific and technical improvements and gradually failed in their objective to cultivate the most modern skills and attitudes to both agricultural students and farmers.

Tamboli and Nene (2013) Indian agriculture has come a long way from the situation living from ship to mouth to food self-sufficiency. The agricultural sector is getting more complex due to globalization, impact of climate change, entry of corporate sector in agriculture value-chain, expanding demand for processed food, and need for post-harvest technology. India will need rich human capital of highly qualified, motivated, and well-trained agricultural scientists to meet these challenges of 21st century. It is the responsibility of the State Agricultural Universities (SAUs) to provide such human resources.

Anupam Verma (2014), there is a need to build Centres of Excellence for agricultural education and research. The country needs at least four more IARIs, located in the east, south, west and central India, by restructuring some of the major ICAR Institutes to cater the need of developing human resource for cutting-edge research. For example, IIHR, Bangalore can be re-designated as IARI, Bangalore. It should then start post-graduate teaching in all disciplines of agriculture. I understand, there are plans to establish de- novo IARI-like institutions. It needs reconsideration.

3.5 The Approach and Methodology Adopted for the Present Study

3.5.1 Data Base

To meet the study objectives, both primary and secondary data were collected from various stakeholders.

3.5.1.1 Primary data

The data were collected through pre-tested questionnaires from various stakeholders of the study includes students, teachers, Alumni, Agro-based industries, Organization/Institutions.

3.5.1.2 Secondary data

University and Colleges: Admitted and out turn of graduates from all 6 farm Universities of Karnataka for the past 5 years (2010 to 2015). Information on Scholarships and Schemes from different colleges was obtained to know the how many students are benefitted from ICAR, SAU and state government schemes. Information from the university Placement Cell were collected, in order to know the trends in employability of graduates for last five years, growth rate of different sub sectors *viz.*, financial institutions, NGO's, private seed, fertilizer & pesticide companies was calculated to estimate the requirement of man power. The present status of sanction, filled and vacant posts at university level cadre-wise Assist Professors, Associate Professors and Professors in different sectors. Annual reports of past five years (2010-2015) were also collected from all the SAU's.

Developmental Departments: In order to assess the demand and supply of graduates. The sanction, filled and vacant posts in the concerned state departments such as Agriculture, Horticulture, Sericulture, Animal Husbandry and Veterinary, Fisheries, Forestry, Watershed and Women and Child Development Department were obtained by personal interview with officials and annual reports.

Financial Institutions: From nationalized bank information on number of agriculture and allied sector graduates working in the institution was collected by personal interview with the official from state headquarter and also data from Reserve Bank of India (RBI) website. Similarly, information from five private banks was collected.

Agro-based Industries: From the HR office of the agro-based industries, available data and information on total number of graduates recruited in private companies/industries such as seed, fertilizer, pesticide companies, equipment dealers, manufacturers of equipments, seed production, nursery, sugar factories, food processing industries were collected.

3.5.2 Sample Design

3.5.2.1 Sample of Students

Primary data from students were collected through pre-tested questionnaires by random sample survey from students with consideration of representing all farm Universities and degree programmes. Total sample 245 students of which Agriculture-30, Horticulture-22, Veterinary-25, Agril. Engineering-14, Agril. Marketing-16, Bio-Technology-12, Dairy Technology-22, Fishery-28, Food Technology-15, Forestry-26, Rural Home Science-24 and Sericulture-11 were randomly selected for the study. 50% of the sample includes women students (Table 3.2).

Table 3.2: Faculty-wise sample of students selected for the study

S.No.	Discipline	No. of Students		
		Male	Female	Total
1	Agriculture	21	9	30
2	Horticulture	16	6	22
3	Veterinary	8	17	25
4	Agril. Engineering	8	6	14
5	Agril. Marketing	8	8	16
6	Bio-Technology	7	5	12
7	Dairy Technology	12	10	22
8	Fishery	14	14	28
9	Food Technology	6	9	15
10	Forestry	10	16	26
11	Rural Home Science	8	16	24
12	Sericulture	5	6	11
	Total	**123**	**122**	**245**

Table 3.3: Sample of teachers selected for the study

S.No.	Discipline	No. of Teachers		
		Asst. Prof.	Assoc. Prof.	Professor
1	Agriculture	2	2	2
2	Horticulture	1	1	2
3	Veterinary	1	1	2
4	Agril. Engineering	1	1	2
5	Agril. Marketing	1	1	2
6	Bio-Technology	1	1	1
7	Dairy Technology	-	1	2
8	Fishery	1	2	2
9	Food Technology	1	1	2
10	Forestry	1	1	3
11	Rural Home Science	1	1	3
12	Sericulture	-	1	2
	Total	**11**	**14**	**25**

Table 3.4: Sample of Alumni selected for the study

S.No.	Nature of Work	Number
1	Employed on salary	22
2	Self-employed	4
3	Looking for work	4
	Total	30

3.5.2.2 Teachers

A sample of 50 faculty members (Assistant Professors-13, Associate Professors-11 and Professors-26) and discussed with education experts of SAU/ ICAR regarding and opportunities after completion of degree, quality of agriculture education system, opinion about students and demand & supply of graduates in the states (Table 3.3).

3.5.2.3 Alumni

In addition to the student's survey, feedback on the adequacy of the education received in the SAUs, securing jobs and handling the jobs, *etc.* were obtained from alumni of recent years. The total number of alumni are 30, including who are employed on salary-22 in different sub-sectors in agriculture and allied sciences, self-employed-4 and looking for job-4 (Table 3.4).

3.5.2.4 Agro-based industries

Sample of 30 agro-based endeavours such as input dealers (9), equipment dealers (3), manufacturers of equipments (3), seed production (8), nursery (3), sugar factories (4), *etc.* were interviewed to get information related to quality of graduates *viz.*, employability skills, working nature, technical knowledge and also demand of graduates such as total employment in the organization at present and future need. Focus Group Discussions were organized with various stakeholders and also from heads of the industries, requested them to give some feedback in the form of comments on graduates and suggestions to improve quality of agricultural education.

3.5.2.5 Organization/ Institutions

From head of various organization/ institutions such as financial institutions, developmental departments, NGO's where the agriculture graduates are working was collected to know the quality of agriculture graduates in terms of employability skills and technical knowledge.

3.5.3 Analysis

Descriptive statistics and percentages were used for analysis of data and draw a meaningful conclusion. The regression analysis was used for estimating the compound annual growth rate (CAGR) of intake and out turn of students, sector growth of agricultural and allied sectors.

Estimation of future requirements of skilled human resource across agriculture and sub-sector was estimated on the basis of demand using area based /services requirement approach. System dynamic model and causal loop model were used for the projection of future human resources in agriculture and allied sectors.

In the light of the project objectives, this study has made an modest attempt to make a forecast of the number of graduates, post-graduates and doctorates likely to be available as well as requirements over the next ten years, and also assess the existing skill gaps and further requirement of skilled manpower.

The base year for the projections has been taken as 2014-15, the year of the commencement of the project. The forecast horizon is taken as 2019-20 to 2024-25.

3.5.3.1 Demand Forecast

Constraints of data availability dictate the choice of methodology that can be adopted for demand forecasting. Keeping in view the massive and diverse nature of the system of human capital utilization in agricultural and allied sectors and the data availability constraints, the system dynamic modal and causal loop modal has generally been adopted in the present project for obtaining quantitative forecasts, which covers the needs of the entire economy. This is supplemented by use of normative approach/ area service approach, whereever necessary.

The projections have been derived through the following steps:

1. Important sub-sectors have been identified in each sector of the agriculture and allied sectors (such as agriculture, horticulture, veterinary, *etc.*). Such sub-sectors, for instance, include government services, finance, processing industry, research and education, input industry, *etc.*
2. The targeted developments (such as growth rates) in each of these sub-sectors are considered as specified in plan and other official

documents or, where such targets are not available, have been derived on the basis of growth in the recent past and projected using linear regression.

3. Total employment in each sub-sector in the base year (2014-15) has been estimated, if not available from the existing sources, using either normative approach or trend analysis using the data from survey and placement cells information.

4. The methods were differing from sub-sector to sub-sector depending on the availability of information. A sample of 10 agro based endeavours such as input dealers, equipment dealers, manufacturers of equipments, seed production, nursery and agriculture related enterprises including departments of agriculture, horticulture, forestry and other line departments, commercial banks, co-operative banks, farmers' service societies, private extension, to assess the potential for employment.

The following methods were used :

a) In the case of some sub-sectors, **data on total employment is available** from other sources, for 2014-15 year on the basis of trend analysis. Such sub-sectors are, **for example,**

Sub-sector	Source	Nature of data
Financial Institutions	Banking Statistics of Reserve Bank of India, for Karnataka	No. of officers and others, each year latest being 2014-15
Different processing industries (Dairy, agro-, fruit, meat, fish, *etc.*	Annual Survey of Industries, Annual reports	Total employment, number of factories, each year, latest being 2014-15
Input industries like fertilizers, pesticides, agricultural equipment, animal feed, *etc.*	-do-	-do-
Output absorbing industries like, pharmaceuticals, paper, wood processing , *etc.*	-do-	-do-
Government departments dealing with the sub-sector	Websites of the respective departments	Total employment, classified by posts
Teaching staff in the SAUs	Institutional schedules or annual report from SAUs	Data on teachers by qualification and field 2014-15

b) For some sub-sectors, data for total employment is not available, but **only on the number of units** in the sub-sector. **Examples** of such subsectors are:

In these cases, the total number of units in 2014-15 will be estimate first. Employment per unit will be obtained from Establishment Survey and total employment in the sub-sector is obtained by multiplying the number of units by the per unit employment.

c) In some cases, per unit employment could not be estimated from Establishment Survey due to data problems. In such cases, norms available from various sources will be of use. For example, in the fisheries sector, such norms as are available from CIFE, were used to estimate the employment in fish hatcheries, *etc.*

d) The total employment has been broken down into educational requirements of stock on the basis of primary data specially collected in the project, as outlined later in this chapter. It is generally assumed that the current pattern of employment holds good for future years also. However, wherever deemed necessary on the basis of expert views expressed in Focus Group Discussions, a judgmental approach has been followed to slightly modify the proportions.

e) The stock projections have been converted to flow projections. The flow projections consist of annual increment in the stock.

f) The replacement needs (due to factors such as retirements, deaths, migration, *etc.*) taken as 3 per cent of the previous year's stock (Rama Rao, *et al.*, 2011).

g) The requirements of higher education takes into account the fact that the alumni at any level do not all enter the labour market straightaway and a significant proportion of them pursue higher education and become available to economic activity only after certain number of years (similar as that of the method, followed by Rama Rao *et al.*, 2011).

In addition, qualitative aspects of human capital needs have been captured through Focus Group Discussions with various stakeholders, experts and opinion surveys of employed and unemployed agricultural human resources.

3.5.3.2 Forecast of Supply

The basis for assessing supply of agricultural human resources is the annual institutional output over the years. In the absence of a comprehensive data base on output from agricultural universities at one place, special efforts had been made to collect the relevant data from all, the all SAUs in the state to assess the levels of output for various years besides taking data from the National Information System on Agricultural Education Network in India (NISAGENET) maintained by Indian Agricultural Statistics Research Institute (IASRI), New Delhi for the estimation of demand and supply of agriculture and allied sector graduates in the state. The system dynamic modal was used to forecast future requirements.

3.5.4 Assumptions

- Due to development interventions of the Government of India in terms of various flagship programmes like national horticulture mission, watershed development, Jandhan yojan, RKVY, *etc.*, as well as other schemes, there would be a need for professionally qualified personnel.
- Presently many agriculture related positions in private sector, personnel from general stream are working. Gradually professionally qualified persons will be hired in all such occupations.
- To meet the target of agriculture growth (4%) and doubling of farmers income, government will adopt pro-active policies and their implementation.
- The expansion of allied sectors of agriculture such as horticulture, dairy, poultry, etc., will continue the pace of development, which would have positive impact upon human capital needs.
- The vacant positions will be filled up at a faster rate to meet the needs of agriculture and allied sectors.
- With rural enterprise development, technological innovations, *etc.*, qualified persons would be required to cater to the needs of farmers.
- Net inflow and outflow of agriculture and allied sector graduates from Karnataka to other states is assumed to be the same.

❑❑❑

Chapter **4**

Human Resource Assessment in Agriculture Sector

4.1.1 Introduction

There are strong direct relationships between agricultural productivity, hunger, poverty, and sustainability. Moreover, the higher the proportion of the rural population that obtains its income solely from subsistence farming (without the benefit of pro-poor technologies and access to markets), the higher the incidence of malnutrition. In order to meet the needs of growing population in the face of shrinking availability of land for cultivation, impact of climatic changes and vagaries of monsoons, response fatigue of land to standard agricultural inputs, decelerating productivity and other factors have led to adopting a multi-pronged approach to boosting agricultural production. The overall strategy includes diversification into promising sub-sectors like horticulture, dairying, fisheries, food processing, *etc.*, but at the same time, raising productivities in the crop sector through greater attention to need-based research, emphasis on transfer of technologies through better extension, development of high-yielding seeds and their distribution, efficient water management, agricultural investments and other measures to raise the production of crops like cereals, pulses, oilseeds, sugarcane, *etc*, essential for tackling the food security concerns, particularly for the poor.

4.1.2 Education in Agriculture Science

Soon after independence, India followed the path of science, which led to the growth of agriculture reaping rich dividends. National Agricultural Research System (NARS), consisting of ICAR Institutes and State Agricultural Universities, has played a pivotal role in the accomplishment in the field of agriculture. The green revolution, with its impressive social and economic impact, would not have been possible without the availability of technically qualified human resource.

Thus, agricultural education is placed in the forefront for building a scientific manpower base. Similarly, in Karnataka first Agriculture University was established during May 25, 1963 at Bangaluru with two colleges in Bangalore and Dharwad. Later University of Agricultural Sciences, Dharwad came into existence during October 1, 1986. Afterwards one more Agriculture University as UAS, Raichur was established in the year 2009 with four colleges. Recently University of Agricultural and Horticultural Sciences, Shimoga was established in September 21, 2012 with five (Agricultural and Horticultural) colleges.

4.1.3 Supply of Graduates

Agricultural education, research and extension have been accorded fairly high priority and elaborate institutional set-up has been established over the years. List of colleges providing education in agriculture science is given in Annexure-1.

There are nine colleges providing education in agriculture (*i.e.*, degree in Agriculture Sciences) and intake and out turn of graduates in various agriculture colleges in Karnataka (Table 4.1.1). The annual admitted strength of students in agriculture in four state Agricultural Universities is 1377 in diploma, 1186 at under-graduate level, 658 at post-graduate level and 136 at doctoral level during 2014-15. Whereas, the annual passed out strength of students in agriculture is 1122 in diploma, 955 at under-graduate level, 637 at post-graduate level and 126 at doctoral level during 2014-15. In the past five years gradual increase in the students intake and out turn 1478 and 1160 in 2010- 11 to 1980 and 1718 in 2014-15 with CAGR 6.77 % and 10.92 %, respectively in UG and above.

Table 4.1.1: Students Admitted and Passed in Agriculture Courses during 2010-11 to 2014-15

	Intake				Out turn				UG and above	
Year	**Diploma**	**UG**	**PG**	**Ph.D.**	**Diploma**	**UG**	**PG**	**Ph.D.**	**Adm**	**Pass**
2010-11	233	941	416	121	223	614	460	86	**1478**	**1160**
2011-12	635	1127	443	144	565	590	413	99	**1714**	**1102**
2012-13	569	1196	536	142	548	674	448	92	**1874**	**1214**
2013-14	468	1074	615	149	548	780	507	129	**1838**	**1416**
2014-15	1377	1186	658	136	1122	955	637	126	**1980**	**1718**
CAGR (%)	**38.39**	**4.23**	**13.26**	**2.71**	**37.72**	**12.33**	**8.94**	**10.83**	**6.77**	**10.92**

4.1.4 Areas of Placement for Agriculture Graduates

A number of sub-sectors of agriculture requires well-qualified human resources to attain maximum growth of the agriculture sector. The skilled human resources in agriculture are mainly required in research and education and extension of suitable technologies to the farmers. Distribution and sale of inputs like seeds, fertilizers, farm equipment and pesticides, production of crops, agricultural credit and crop insurance services, pricing and marketing, post-harvest storage, quality control, and processing and wholesale and retail sales to the consumer. Agri-businesses, agri-clinics, agro-service centres and other modes of self-employment also have a large potential to employ human resources in agriculture science (Annexure-2).

4.1.5 Current Stock of Human Resources in Agriculture Science

There is no direct source of data on the number of active agriculture science human resources in the state. As such, the current stock can only be estimated, as in the case of other agriculture related disciplines, indirectly by cumulating the out turn of graduates, post-graduates and doctorates in agriculture sciences over past five years (as many as the active working life of a graduate, post-graduate or doctor) and adjust for attrition due to retirement, migration and such causes.

NAARM, in its previous study on assessment of future human capital requirements in agriculture and allied sectors in India, had estimated the annual supplies of total stock of graduates and above in agricultural (crop) sciences is about 15,949 in 2010 from all over India. These stocks were projected to grow to 25,284 by the year 2020.

An effort was made to estimate the stock of agriculture graduates from annual supply data from all agricultural universities. The stock from supply side is estimated from the out turn of graduates in the past 5 years (details given separately in Annexure-4.2). The total stock of graduates and above in agriculture science, estimated from annual supplies is about 1718 in 2014-15.

4.5 Demand Projections

4.5.1 Development Departments

State developmental departments have major responsibility to formulation and implementation of policies and programmes for agricultural development through the Ministry of Agriculture in the State. Large number of vacancies of extension workers in the state Agriculture Department, and other Govt. Departments are the biggest employers of formally qualified agricultural human resources. The network of agricultural offices in the state, the primary purpose of which is to provide various agriculture-related services down to the farmers and stimulate development of agriculture in the state, extends to divisions, districts, blocks and sub-divisions of blocks in the state. Staff position details of agriculture human resources in the state are given in Table 4.1.2.

The number of agricultural officers post filled in the state has been estimated at 3512. There are about 32.54 per cent of total sanctioned posts of officers are vacant in the department of agriculture (Table 4.1.2). In addition, in the state there are autonomous organizations, which also recruit the agriculture graduates.

Organisation survey revealed that approximately more than half of the officers in the government were graduates (67.33 %), post-graduates (20.00 %) and doctoral (2.00 %). It is assumed that these proportions hold good till 2025.

Detailed projected stock demand in development sector is shown in Annexure-4.2.1 These projections estimate that the stock requirement in 2014-15 is about 1694. This would increase to 2162 and 2759 by 2019-20 and 2024-25, respectively. Thus, government agriculture services would need about 100 per annum during 2015 to 2025.

4.5.2 State Agriculture Universities

The main mandate of the agriculture universities are education, research and extension. The number of trained human recourses is required in the SAU's to carry out various activities in teaching research and extension. Due to unavailability of information related to sanctioned, filled and vacant positions by discipline. Demand of graduates and above in teaching, research and extension are discussed together. In SAU's education system, number of skilled manpower require for effective teaching to the students. The universities recruit the various trained agriculture personals in the cadre of assistant professors, associate and professor cadre. The agriculture extension work by the SAU's is carried out by a number of Krishi Vigyan Kendras and extension service centre in the state. Krishi Vigyan Kendras (KVKs) attached to various SAUs constitute the principal extension organization in the country for transferring farm technology and training farmers. There were 32 KVKs in Karnataka as on 2015 (www.icar.org.kvk karnataka). On an average, a KVK has 3 to 8 (average of 5) professional agricultural scientists (Project Coordinator and Subject Matter Specialists they include horticulturalists, fisheries, animal science professionals as well). The projections under extension activity include only KVKs, agri-clinics and other NGOs, which are primarily in extension activities. The human resources requirements of other providers of extension services namely government, input industries and retail, *etc.* have been included in the major activities of these respective sectors. It may not be possible to separate the number of staff specifically engaged in extension activities in these sectors.

Research is being carried out in the universities through projects operated and administered by the Directorates of Research and their sub-centres. This activity is not considered separately for projections, but is taken into account while assessing the SAUs requirements under Academic sector. In the private sector, the research is mostly related to seed development and the needs are considered under that sub-sector. In view of the limits on growth in the area cultivated for field crops in future due to factors like diversification into other sub-sectors like horticulture, and the claims of urbanization for housing and other infrastructure development, future growth in the crop sector has to come from enhancement of productivity, which demands need-based research. Research in crop sciences takes place both in the SAUs and private institutions. The SAUs (including all teaching, research and extension) had a combined sanctioned strength of 1429 faculties, of which 532 positions (37.23 %) were vacant as on March 31, 2015 (Table.4.1.2). Considering past experience, the number of sanctioned posts is not likely to increase significantly, and any expansion may only reduce the current gap of vacant posts. It may be assumed that the gap may at best be reduced from the current 37.23 per cent to 20 per cent by 2025.

Detailed projected stock demand in SAU's is shown in Annexure-4.2.2. These projections estimate that the stock requirement in 2014-15 is about 532. This would increase to 639 and 867 by 2019-20 and 2024-25, respectively. Thus, SAU's including Teaching, Research and Education; need about 80 per annum during 2015 to 2025.

Table 4.1.2: Staff positions in the developmental departments and SAU's as on March 31, 2015

Organisation	Sanctioned	Filled	Vacant	Vacancies %
SAUs	1429	897	532	37.23
Devp. Department	5206	3512	1694	32.54
Total	**6635**	**4409**	**2226**	**33.55**
SAUs- Agriculture (UAS -B, D, R and S)				
Devt. Department- Agriculture				

4.5.3 Financial Institutions

NAARM in its previous study estimated that, about 4 per cent of the total employment (or, 10 % of the officers in banks, as officer strength is 40% of the total employment) in banks comprised graduates and post-graduates in agriculture and allied sciences. About 50 per cent of them were graduates; 40 per cent post-graduate; doctorates forming a mere 5 per cent, the rest being diploma/certificate holders in agri- related fields. Employment of officers in public sector banks has been growing at an annual rate of 3.7 per cent during the last decade (2000-01 to 20010-11) - the rate being 5 per cent in urban and metropolitan branches and 1.4 per cent in the semi-urban and rural branches of the banks. Two projections of total employment of officers in banks have been made - a low projection with 1.5 per cent per annum growth and a high projection with 3.7 per cent annual growth leading to estimated 5.1 lakh officers under the low projection and 6.4 lakhs under the high projection. As 10 per cent of the officers are with agri-related qualifications of graduates and above, the number of such personnel has been estimated as 44,000 and 56,000 in 2020 under the two projections, respectively (Rama Rao *et al.,* 2010).

The financial institutions such as nationalized and private banks are major employers for agriculture graduates. Presently, agriculture graduates are recruiting through banking exam conducted by Institution of Banking and Personal Selection (IBPS). Whereas, private banks by direct interviews. Financial institution is currently growing at 5.0 per cent and this trend is likely to continue. By looking into past five years of data obtained from placement cells, all the SAU's the employment pattern of graduates in the financial institutions revealed that the growth rate (CAGR) is about 5.60 % from 2010 to 2015. The details of financial institution recruitments through placement cell are given in Annexure-4.

Detailed stock projections for financial institutions are given in Annexure-4.2.3. As 10 per cent of the officers are with qualifications of agriculture and allied sector related graduates and above, the number of such personnel has been estimated as 350 in 2015 and 604 in 2025. Thus, the requirement is about 25 per annum.

4.5.4 Seed Sector

Seed is the basic and most critical input for sustainable agriculture. The response of all other inputs depends on quality of seeds to a large extent. It is estimated that the direct contribution of quality seed alone to the total production is about 15–20 % depending upon the crop and it can be further raised upto 45% with efficient management of other inputs. Adequate availability of quality seeds is a particular challenge for farmers and productivity depends crucially on timely sowing. The seed system must be capable of providing seeds of contingency or alternative crops during sowing time. The responsibility for seed production is shared between ICAR research institutions, SAUs, National Seeds Corporation (NSC), State Farms Corporation of India (SFCI), State Seeds Corporations, KVKs, State Department of Agriculture, as well as private sector with necessary infrastructure and expertise. Apart from NSC, SFCI, State Seed Corporations and other institutions in the public sector, and about various private sector establishments (Annual Report Seed Association of India) are in the field of seed production. The distribution of seeds is undertaken through a number of channels *i.e.* departmental outlets at block and village level, cooperatives, outlets of seed corporations, private dealers, *etc.* There are also about 7 seed testing laboratories in the state in different districts and SAU's.

Production and supply of Breeder Seeds of different crops is being done by the State Agricultural Universities and Indian Council of Agricultural Research Institutes (ICAR). These organizations produce and supply Breeder Seeds required for the State based on the advance indents placed by the Department of Agriculture. The Seed Association of India is the Nodal agency for supply of breeder seeds to the private seed producing agencies. Certified Seed production and distribution in the State is being done by Karnataka State Seeds Corporation, National Seeds Corporation, Karnataka Co-operative Oilseed Growers Federation, State Agricultural Universities and number of private seed supplying agencies. Seed producing and supplying agencies fix the procurement and selling rates of various seeds taking into account, procurement rates, production costs, certification, processing, packaging and other overhead charges. Totally 4.36 lakh quintal seeds have been distributed and Rs. 7420.46 lakh subsidy amount is utilized for this purpose. Totally 17.53 lakh farmers have been benefitted, during Rabi/Summer, 2015-16, it was programmed to distribute seeds of 12 crop seeds under subsidy. Up to the end of December 2015, totally 3.24 lakh qtl. seeds have been distributed and Rs. 7722.51 lakh subsidy amount is utilized for this purpose and 6.81 lakh farmers have been benefitted.

The production of Breeder and Foundation seed and availability of certified seed has been increasing in past few years and the growth has been particularly fast in recent years. Actual production and distribution of certified seed is about 10,23,172 qtl. and 12,70,340 qtl. in 2012-13 and 11,65,250 qtl. and 15,75,320 qtl. in 2015-16, respectively (Anon., 2017).

The private sector has started to play a significant role in the seed industry over the last few years. At present, the number of companies engaged in seed production or seed trade is > 200. However, the main focus of private seed companies has been on the high value low volume seeds and market for low value high volume seeds of cereals, pulses and oilseeds is still dominated by the public sector seed corporations. Private sector, today accounts for about 70 per cent of the market in terms of value while public sector has greater share in terms of volume sales (Vinay *et al.*, 2010). Several large multinational seed firms such as Cargill, Monsanto, Bayer Crop Science, Advanta, Dupont are now operating in Karnataka, with a focus on bio-technology-based research. A number of Indian companies like Indo-American Seeds also make a significant contribution to the seed industry. Private industry is largely concentrated around Bengaluru, Haveri and Koppal districts.

The seed sector has been growing in recent years to 10 to 12 per cent per year. Discussions with major seed industries indicated that the sector needs rapid growth of around 15 per cent in future. By looking into past five years, Placement Cell interview details of all the SAU's the employment pattern of graduates in the seed industry ranges from 10 to 53 during 2010 to 2015, respectively. The growth rate (CAGR) is about 31.70 % from 2010 to 2015. The details of seed companies' recruitments through placement cell are given in Annexure-4. Projections of demand for the future years have been obtained assuming a better rate of 10 per cent per annum.

The detailed projections of requirements for the seed sector are shown in Annexure-4.2.4. The stock requirement of human capital would grow from 90 in 2014-15 to about 233 by 2025, i.e. about 10 per annum during the period 2015 to 2025.

5.5.5 Fertilizers

Fertilizers are one of the important inputs in agriculture for crop production and have been major contributor to the attainment of food security. The Government of Karnataka projected a requirement of 22.70 lakh tonnes of different grades of fertilizers consisting DAP 4.00 lakh tonnes, MOP 2.50 lakh tonnes, NPK 6.50 lakh tonnes, Urea 8.50 lakh tonnes and SSP 1.20 lakh tonnes for Kharif 2015. At the end of Kharif season 22.45 lakh tonnes of Fertilizer has been distributed consisting DAP 3.68 lakh tonnes, MOP 1.62 lakh tonnes, NPK 6.72 lakh tonnes, Urea 8.62 lakh tonnes and others 1.81 lakh tonnes fertilizers has been utilized. For Rabi 2015-16, the State Government has projected a requirement of 15 lakh tonnes of different fertilizers consisting of DAP 2.00 lakh tonnes, MOP 1.50 lakh tonnes, NPK 5.50 lakh tonnes, Urea 5.50 lakh tonnes and SSP 0.50 lakh tonnes. At the end of Rabi season (upto December 2015), 6.60 lakh tonnes of fertilizer has been distributed and utilized consisting of DAP 0.60 lakh tonnes, MOP 0.34 lakh tonnes, NPK 2.02 lakh tonnes, Urea 3.44 lakh tonnes and SSP 0.52 lakh tonnes. However, data on fertilizer consumption and HYV Coverage in Karnataka during the last few decades indicate almost the increase in consumption of fertilizers, which now account for 8.5 per cent.

The fertilizer industry is currently growing at 8 per cent and this trend is likely to continue. By looking into past five years Placement Cell interview details of all the SAU's the employment pattern of graduates in the fertilizer industry ranges from 24 to 56 during 2010 to 2015, respectively. The growth rate (CAGR) is about 23.20 % from 2010 to 2015. The details of fertilizer companies' recruitments through placement cell are given in Annexure-4.

Detailed projections are given in Annexure-4.2.5. The requirement of human capital would grow from 70 in 2010-11 to about 114 by 2025. The requirement during 2015- 25 is about 10 per annum.

4.5.6 Pesticides

In order to protect the crop against pest and diseases, and ensure the secured food production pesticides are plays major role. The similar trend as observed in fertilizer sector was observed here also with respect of employability, and recruitment through placement cell. The pesticides industry is currently growing at 4.2 per cent and this trend is likely to continue. By looking into past five years Placement Cell interview details of all the SAU's, the employment pattern of graduates in the pesticides industry ranges from 14 to 81 during 2010 to 2015, respectively. The growth rate (CAGR) is about 26.3 % from 2010-2015. The details of pesticides companies' recruitments through placement cell are given in Annexure-4.

Detailed projections of requirements in the pesticides sector are given in Annexure-4.2.6. The requirement of human capital would grow from 50 in 2014-15 to about 67 by 2025, i.e., about 5 per annum.

4.5.7 Entrepreneurship

For people interested in entrepreneurship, there is sufficient technical guidance, financial assistance and subsidies provided by the NABARD, Agri-clinic and Agri Business Centre and other nationalized banks. NABARD gives financial support to the candidate having B.Sc. degree, for agriculture-related self-employment enterprises. The main areas for entrepreneurs in Karnataka are Integrated Farming System, Soil testing centres, Seed production, Crop disease diagnostic centre, Consultancy services and Establishment of agri-clinics, *etc.* In order to know the number of graduates who are self-employed, the information were collected from alumni association and also disused with experts. There is about less than 1 per cent agriculture graduates are self-employed due to various reason details are discussed in the Chapter12.

4.5.8 NGO's

Graduates who are oriented more towards community development and social upliftment of agriculture, dependent population, they can join Non-Governmental Organization. By looking into past five years Placement Cell interview details of all the SAU's, the employment pattern of graduates in the NGO's range were from 5 to 24 during 2010 to 2015, respectively. The growth rate (CAGR) is about 1.35 % from 2010-2015. The details of NGO's recruitments through placement cell are given in Annexure-4.

Detailed projections of requirements in the NGO's have been shown in Annexure-4.2.7. The stock requirement of human capital would grow from 50 in 2014-15 to about 67 by 2024-25, *i.e.* additional requirement of 2 graduates per annum.

4.5.9 Agro-processing Sector

Processing of agricultural crop produce takes place both in informal as well as in the organized sectors. Apart from the traditional processes of milling of cereals and pulses, extraction of oils from oilseeds, extraction and refining of sugar and various tobacco-related activities, manufacture of various value-added food products of quality has been growing rapidly. Various agro-processing industries employ a large number of agricultural human resources. About 5 per cent of the graduates and post-graduates found employment in agro-industries (establishment survey).

Annual Survey of Industries (ASI) data for agro-industries (other than horticultural, fishery, dairy and animal husbandry-related industries) cover manufacturing of sugar and sugar products, tobacco products and various other food products like grain mill products, *etc.* of these sugar (including co-operative sugar mills) are the very large employers with the sugar industry employing about 50 graduates.

As per the Establishment Survey, 5 per cent of the total employment are those with agri-skills, about 70 per cent of whom are graduates and 30 per cent are post-graduates. On this basis, the requirement of the processing sector by 2025 has been projected at a total of 81 Agriculture Graduates (including post-graduates).

Detailed stock projections based on these calculations for agro-processing sector are given in Annexure-4.2.8. The additional requirement during 2014-15 to 2024-25 would be about 5 per annum.

4.5.9 Other Sectors

There are a number of other societies, boards and institutions also recruit the agriculture such as, ICT, Commodity boards, management institutions, business agencies and foundations at national and international level. Placement Cell interview details of all the SAU's the employment pattern of graduates in the other sectors ranges from 31 to 34 during 2010 to 2015, respectively. The growth rate (CAGR) is about 13.02 % from 2010-2015, Annexure 4.2.9. The details of other sectors recruitments through placement cell are given in Annexure-4.

Detailed projections of requirements in the other sectors have been shown in Annexure-4.2.9 The stock requirement of human capital would grow from 200 in 2014-15 to about 326 by 2025, *i.e.* addition requirement of 3 graduates per annum.

The overall growth rate in the employment in various private sectors is about 17.90 % as per placement cell information from 2011 to 2015.

4.6 Summary of Demand Projections

The projections of stocks of agriculture human resources required to meet the needs of various segments of the economy and the consequential flows required from the educational system are summarized below:

4.6.1 Stock Projections

The overall stocks of Agriculture human resources required to meet the requirements of various segments of crop sciences have been added to give projections upto 2024-25. Three sets of projections have been worked out as explained in the segment- level projections – one a high projection, a low projection and an average. These projections are given in **Annexure-4.** The corresponding annual outputs required from the educational system to lead to these stocks are given in Annexure-4.2.

Considering the current supply and experts opinions on graduates aspirations and employee organizations capacity to absorb graduates, the average variant projection is considered desirable. The overall stock as per the average scenario is given in **Table-4.1.3.** The stock of diploma holders would grow from 77 in 2015 to about 125 in 2025. On the other hand, the requirement for graduates and above would grow from about 3931 in 2015 to about 6403 in 2025, *i.e.*, the average addition to the stock during the period 2015 to 2025 is about 210 per annum.

Table 4.1.3: Overall Projections of Requirements Stocks of Agriculture Human Resources

Year	Diploma	UG	PG	PhD	All UG & above
2014-15	77	1973	1243	657	3931
2024-25	1336	3309	2014	1056	6495

4.7 Demand – Supply Gap in Stock

In 2014-15, the actual stock from supply is 1718. Assuming out turn levels will grow at the same growth rate as observed during the last five years (*i.e.*, 10.92 per cent during 2010-11 to 2014-15), the estimated demand–supply scenario in 2025 will be:

Supply: 4902

Demand: 6495

Gap: 1591 (24.52 % of the demand)

4.8 Flow Projections

The overall annual flow required in all sub-sectors of agriculture science is given in Annexure-4.1.1 and summary in Fig. 4.1.

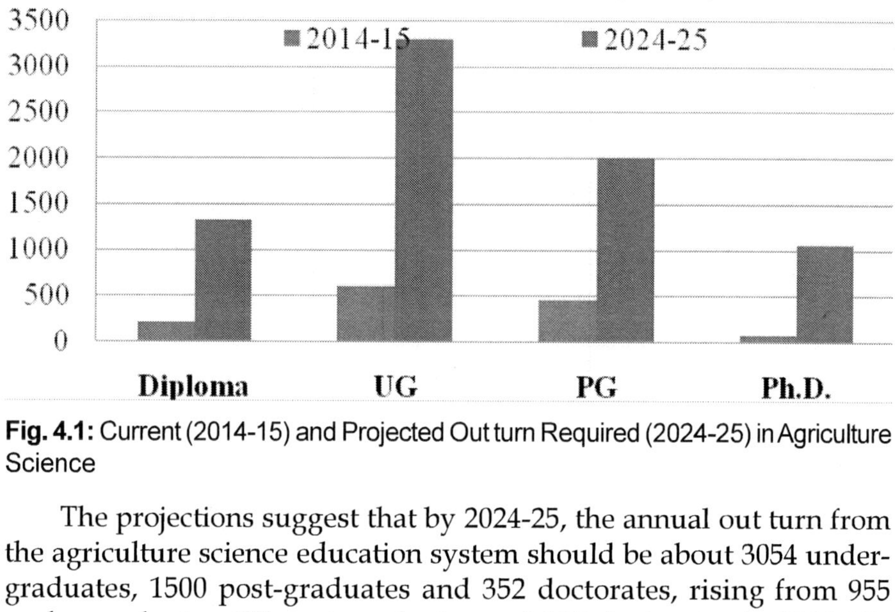

Fig. 4.1: Current (2014-15) and Projected Out turn Required (2024-25) in Agriculture Science

The projections suggest that by 2024-25, the annual out turn from the agriculture science education system should be about 3054 under-graduates, 1500 post-graduates and 352 doctorates, rising from 955 under-graduates 637 post-graduates and 126 doctorates in 2014-15.

4.2 Human Capital Assessment in Agricultural-Marketing and Co-operation

4.2.1. Education in Agriculture Marketing and Co-operation

The term "Agri-marketing" refers to the entire range of business initiatives relating to food production, and includes contract farming, seed supply, fertilizers and pesticides, farm machinery, and post-harvest activities like storage and transport, wholesale and retail trade of agricultural commodities, processing and retail sales. Apart from activities relating to agricultural inputs like seeds, fertilizers and pesticides already covered in the above paragraphs, modern retail trade in agricultural commodities is an emerging area, which has potential to employ agricultural human resources.

In order to supply the skilled human resource for field of Agriculture Marketing and Co-operation in Karnataka, the degree program of Agricultural Marketing and Co-operation was stated in the year 1976 in erstwhile UAS, Bengaluru at College of Agriculture GKVK and College of Agriculture, Dharwad simultaneously with 25 intake capacity, each at graduation level. Later on PG programmes were stated during 1994-95 at College of Agriculture GKVK with intake capacity of 5.

Modern agriculture practices are increasingly turning out to be knowledge-based and hence, gaining expertise in them is not an easy task for many of our rural farmers. Education focused on the effective marketing skills to facilitate infrastructure facilities and techniques of preservation, storage, and transportation, *etc.* with a view to reduce the post-harvest losses and ensuring better returns is provided to farmers. Institutions like Indian Council of Agricultural Research and Marketing Research and Information Network (AGMARKNET) are working for the cause of upgradation and dissemination of market intelligence for the rural community.

4.2.2 Supply of Graduates

The out turn of graduates in Agriculture Marketing and Co-operation are 105 (2014-15), among them 69 are under-graduate, 33 are post-graduate and 3 are doctoral degree graduates. Looking into past five years data of intake and out turn of graduates in Agriculture Marketing and Co-operation, the decreasing trend in intake 133 in 2010-11 to 127 in 2014-15 with CAGR at 0.74 %, whereas increasing trend in out turn 39 in 2010-11 to 69 in 2014-15 at under-graduate level with CAGR at 18.46 %. In master's degree, increasing trend existed in the intake and out turn 29 and 26 in 2010-11 to 31 and 33 in 2014-15 with CAGR 0.04 and 5.61 %, respectively. There was steady trend in the intake of doctoral graduates 3 to 3 (2010-11 to 2014-15) with CAGR 5.24 %, whereas increasing trend in outturn of doctoral 2 to 3 (2010-11 to 2014-15) with CAGR 14.13 % (Table 4.2.1).

Table 4.2.1: Intake and Out turn of Graduates in Agricultural Marketing and Co-operation Course during 2010-11 to 2014-15

Year	Intake			Out turn			UG and above	
	UG	PG	Ph.D.	UG	PG	Ph.D.	Adm	Pass
2010-11	133	29	3	39	26	2	**165**	**67**
2011-12	122	33	3	61	28	3	**158**	**92**
2012-13	138	44	4	70	32	4	**186**	**106**
2013-14	144	29	5	106	30	5	**178**	**141**
2014-15	127	31	3	69	33	3	**161**	**105**
CAGR (%)	**0.74**	**0.04**	**5.24**	**18.46**	**5.61**	**14.13**	**0.70**	**14.17**

4.2.3 Areas of Placement for Agriculture Marketing and Co-operation Graduates

A number of well-qualified Agricultural Marketing and Co-operation human resources are required for the various sectors of the economy for both public and private. Agricultural Marketing and Co-operation graduates mainly required for the effective distribution and sale of inputs, farm equipment and agricultural credit and crop insurance services, pricing & marketing and wholesale & retail sales to the consumer, agri-businesses, agro-service and other services.

4.2.4 Current Stock of Human Resources in Agricultural Marketing and Co-operation

There is no direct source of data on the number of active Agricultural Marketing and Co-operation human resources. Hence, the same procedure was followed as in case of agriculture sector.

An effort was made to estimate the stock of Agriculture Marketing and Co-operation graduates from annual supply data from two colleges. The stock from supply side is estimated from the out turn of graduates in the past 5 years (details given separately in Annexure-4.1.2). The estimated total stock of graduates and above in Agriculture Marketing and Co-operation, is about 105 in 2014-15.

4.2.5 Demand Projections

The major employer of Agriculture Marketing and Co-operation graduates are the Department of Agril. Marketing. The state has 157

Agricultural Produce Market Committees (APMC's) to facilitate and regulate the marketing of agricultural commodities. The Department of Agricultural Marketing supervises and guides the working of APMCs to implement the provisions of the Act and Rules of the department. The aim of the department is to develop and regulate participative, transparent and scientific agricultural marketing system with adequate infrastructure and user friendly e-initiatives in the state. Since department has provided infrastructure facilities for better marketing of the commodities, the value of total arrivals in the regulated markets in the state has increased from Rs.33,165 crore during 2013-14 to Rs. 34,116 crore during 2014-15. The challenges of the Department are to provide additional required infrastructure for trade in the markets, to ensure transparency in the markets, to modernize the markets, and to ensure competitive price to the farmers produce. Online trading system is introduced in 103 markets during 2015-16 upto December, 9,08,507 farmers have registered their names so far to participating at Unified Marketing Platform. In this regard, sufficient number of human recourses is need for various purposes in agriculture marketing sector.

Agricultural human resources requirements of the retailing sector have been perceived on the basis of focus group discussions and taking into account the present status of the sector. It has been pointed out that the sector has tremendous potential for growth and employing agri-professionals. Many of the public sectors organizations and private companies are recruiting agriculture marketing and co-operation graduates in Karnataka, *viz.*, Department of Agricultural Marketing and Department of Co-operation, Govt. of Karnataka having 1552 vacancies (62.10 %), in SAU's 20 vacancies (25.00 %), whereas in private around 40 graduates are required per annum. Detailed projections of requirements in the other sectors have been shown in Annexure-4.3.1.

Stock projections for the organized agriculture marketing and co-operation sector both in public and private have been shown in Annexure-4.3.1. The stock requirement of human capital would grow from 405 in 2014-15 to about 655 by 2024-25, i.e. on an average required 25 persons per annum.

Table 4.2.2. Staff positions in the developmental departments and SAU's as on March 31, 2015

Organisation	Sanctioned	Filled	Vacant	Vacancies %
SAUs	40	30	10	25.00
Department	499	189	310	62.12
Total	539	219	320	59.37
SAUs- Agriculture (UASB, D, R and S)				
Department : Department of Agricultural Marketing and Department of Co-operation, Govt. of Karnataka				

4.2.6 Summary of Demand Projections

The projections of stocks of agriculture marketing and co-operation human resources required to meet the needs of various segments of the economy and the consequential flows required from the educational system are summarized below:

4.2.6.1 Stock Projections

The overall stocks of Agriculture Marketing and Co-operation human resources required to meet the requirements of various segments of Agriculture Marketing and Co-operation have been added to give projections upto 2024-25. Three sets of projections have been worked out as explained in the segment- level projections – one a high projection, a low projection and an average. These projections are given in Annexure-4.3.1. The corresponding annual outputs required from the educational system to lead to these stocks are given in Annexure-4.3.1.

Considering the current supply and experts opinions on graduates aspirations and employee organizations capacity to absorb graduates, the average variant projection is considered desirable. The overall stock as per the average scenario is given in **Table-4.2.3.** The stock of graduates and above would grow from about 415 in 2014-15 to about 681 in 2024-25, *i.e.*, the average addition to the stock during the period 2015 to 2025 is about 35 per annum.

Table 4.2.3: Overall Projections of Requirements Stocks of Agricultural Marketing and Co-operation Human Resources

Year	UG	PG	PhD	All UG & above
2014-15	243	136	37	415
2024-25	456	221	61	681

4.2.7 Demand – Supply Gap in Stock

In 2014-15, the actual stock from supply is 105. Assuming out turn levels will grow at the same growth rate as observed during the last five years (*i.e.*, 11.04 per cent during 2010-11 to 2014-15), the estimated demand-supply scenario in 2025 will be:

Supply: 305

Demand: 681

Gap: 376 (55.21 % of the demand)

4.2.8 Flow Projections

The overall annual flow required in all sub-sectors of Agriculture Marketing and Co-operation is given in Annexure-4.1.2 and summary in Fig. 4.2.

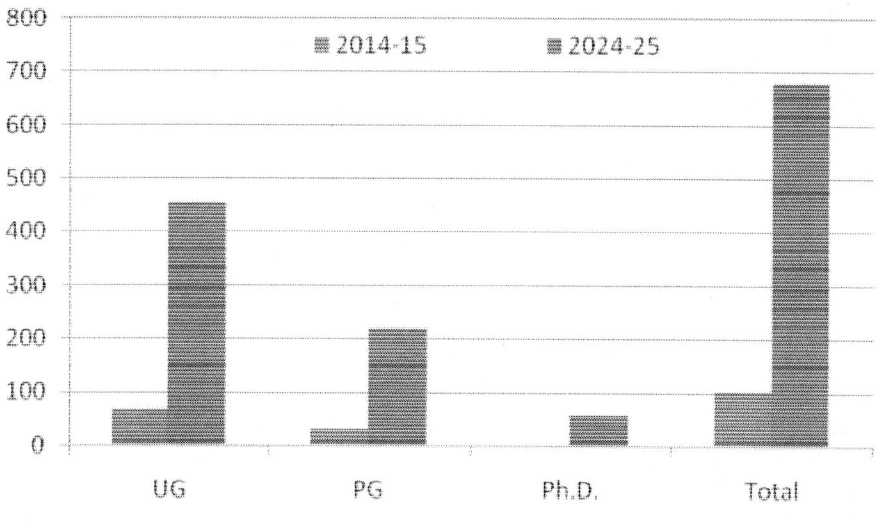

Fig. 4.2: Current (2014-15) and Projected Out turn Required (2024-25) in Agricultural Marketing and Co-operation

The projections suggest that by 2024-25, the annual out turn from the Agriculture Marketing and Co-operation education system should be about 456 under-graduates 221 post-graduates and 61 doctorates, rising from 69 under-graduates 33 post-graduates and 3 doctorates in 2014-15.

4.3 Human Capital Assessment in Agriculture Engineering Sector

Agricultural Engineering is the area of engineering concerned with the design, construction and improvement of farming equipment and machinery. Agricultural engineers integrate technology with farming. For example, they design new and improved farming equipment that may work more efficiently, or perform new tasks. The agricultural engineer plays very important role to make farming sustainable, safe, and environmentally friendly and also find better ways to reduce crop loss from field damage, during handling, sorting, packing and processing. The warehousing of food and fiber are an important part of the agriculture. Hence, in order to make farming sustainable, the trained manpower in the field of agriculture engineering is very much essential. With this background, the education in Agriculture Engineering was started in Karnataka by establishment of Agricultural Engineering Institute in Raichur during 1969, offering the Diploma course in Agricultural Engineering. Later it was upgraded as a college one Agricultural Engineering during 1987 to offer the four years B-Tech (Ag.Engg.) degree programme. Presently, it is only one Agricultural Engineering College in the state but at college of Agriculture UAS, GKVK also offering B.Tech. (Ag.Engg.) degree programme. The PG programmes *i.e.* M.Tech (Ag.Engg.) with specialization in Soil and Water Engineering, Farm Power and Machinery and Agril. Processing and Food Engineering are started recently.

Agricultural engineering is the discipline that applies to engineering science and technology to agricultural production and processing. Agricultural engineering combines the disciplines of animal biology, plant biology, and mechanical, electrical and chemical engineering principles with a knowledge of agricultural principles. It utilises the knowledge of engineering for making agricultural machinery. Importance of engineering inputs for increasing productivity and reducing human drudgery in farming operations is well known in the literature. Agricultural engineering industries have helped the farmers by manufacturing improved farm equipment and processing machinery by adopting suitable designs. In order to achieve efficiency and higher productivity of Indian agriculture, interventions from the agriculture engineering discipline are critical in the form of agricultural mechanization, irrigation and drainage and soil conservation engineering, agricultural processing, energy in agriculture and transfer of technology (Rama Rao *et al.*, 2010-11 NAARM).

4.3.1 Supply of Graduates

During 2014-15, the out turn of graduates in UG and above in Agriculture Engineering from UAS, Bengaluru and UAS, Raichur in Karnataka are 98, among them 74 are under-graduate, 23 are post-graduate and 1 are doctoral degree graduates. There are about 59 Agriculture Engineering diploma holders came out from all the colleges in Karnataka. Looking into past five years data of intake and out turn of graduates in Agriculture Engineering, the increasing trend in intake 116 in 2010-11 to 119 in 2014-15 with CAGR at 0.86 % and out turn was 69 in 2010-11 to 74 in 2014-15 at under-graduate level with CAGR at 7.11 %. In master's degree, increasing trend existed in the intake and out turn 24 and 19 in 2010-11 to 25 and 23 in 2014-15 with CAGR 2.39 and 5.18 %, respectively. Similar, trend in the intake of doctoral 1 to 7 (2010-11 to 2014-15) with CAGR 14.87 % and there was steady trend in outturn of doctoral degree during 2010-11 to 2014-15.

Table 4.3.1: Intake and Out turn of Graduates in Agricultural Engineering Courses during 2010-11 to 2014-15

Year	Intake				Outturn				UG and above	
	Diploma	UG	PG	Ph.D.	Diploma	UG	PG	Ph.D.	Adm	Pass
2010-11	28	116	24	1	26	69	19	1	**141**	**89**
2011-12	48	115	24	2	28	55	23	1	**141**	**79**
2012-13	68	116	26	5	48	63	23	2	**147**	**88**
2013-14	59	119	28	8	68	95	26	2	**155**	**123**
2014-15	59	119	25	7	59	74	23	1	**145**	**98**
CAGR (%)	**18.49**	**0.86**	**2.39**	**14.87**	**28.74**	**7.11**	**5.18**	**7.18**	**1.52**	**6.56**

4.3.3 Areas of Placement for Agricultural Engineering Graduates

A number of well-qualified Agricultural Engineering graduates are required for the various sectors of the economy (public and private). Agricultural Engineering graduates absorbed in farm equipment and processing industry and Micro-Irrigation Companies, Banks, Universities, other public institutions/organizations like RDPR, KSRTC, Administrative posts and self-entrepreneurship (Annexure-2).

4.3.4 Current Stock of Human Resources in Agricultural Engineering

There is no direct source of data on the number of active Agricultural Engineering human resources in the state. Hence, the same procedure was followed as that of agriculture sector. An effort was made to estimate the stock of Agricultural Engineering graduates from annual supply data from different colleges. The stock from supply side is estimated from the out turn of graduates in the past 5 years (Annexure-4.1.3). The total stock of graduates and above in Agricultural Engineering, estimated from annual supplies is about 98 in 2014-15.

4.3.5 Demand Projections

Agricultural Engineering has emerged as one of the most important profession over the past few decades. Advanced farm mechanization, efficient irrigation technologies, post-harvest handling, value addition, *etc* are playing a pivotal role in agricultural renewal and agrarian prosperity.

The major areas, where the agricultural engineers employed, are government departments dealing with soil & water conservation and research and development organisations including private establishments (mainly manufacturing of agricultural machineries) are having significant demand for agricultural engineering human resources in the state. The data related to the staff positions in the Watershed development department, Govt. of Karnataka have been collected (Sanctioned filled and vacant positions) and estimates have been made on that basis. The estimate showed that there is a 37.50 % vacancy in the state. Similarly, for research, extension and teaching data from all the farm agricultural universities in the state has been taken for projections (27.78 % vacancy) Table-4.3.2.

Agricultural engineers requirements in the manufacturing of agricultural machineries, micro-irrigation, and other private sectors have been growing tremendously, there are about 80 graduates are required per annum. Detailed projections of requirements in the other sectors have been shown in Annexure-4.4.1 and 4.1.2.

Stock projections for the organized Agricultural engineering sector both in public and private have been shown in Annexure-4.4.1. The stock requirement of human capital would grow from 150 in 2014-15 to about 415 by 2024-25, i.e. on an average required 30 persons more per annum.

Table 4.3.2: Staff positions in the developmental departments and SAU's as on March 31, 2015

Organisation	Sanctioned	Filled	Vacant	Vacancies %
SAUs	90	65	25	27.78
Department	40	25	15	37.50
Total	**130**	**90**	**40**	**30.77**
SAUs-Agriculture (UASB, D, R and S)				
Department: Watershed Development Department, Govt. of Karnataka				

4.3.6 Summary of Demand Projections

The projections of stocks of agricultural engineering human resources required to meet the needs of various segments of the economy and the consequential flows required from the educational system are summarized below:

4.2.6.1 Stock Projections

The overall stocks of Agricultural engineers required to meet the requirements of various segments of agricultural engineering have been added to give projections upto 2024-25. Three sets of projections have been worked out as explained in the segment- level projections – one a high projection, a low projection and an average. These projections are given in Annexure-4.4. The corresponding annual outputs required from the educational system to lead to these stocks are given in Annexure-4.4.

Considering the current supply and experts opinions on graduates aspirations and employers organizations capacity to absorb graduates, the average variant projection is considered desirable. The overall stock as per the average scenario is given in **Table-4.3.3.** The stock of graduates and above would grow from about 98 in 2014-15 to about 364 in 2024-25, *i.e.,* the average addition to the stock during the period 2015 to 2025 is about 40 per annum.

Table 4.3.3: Overall Projections of Requirements Stocks of Agricultural Engineering Human Resources

Year	Diploma	UG	PG	PhD	All UG & above
2014-15	59	74	23	1	98
2024-25	95	151	160	45	355

4.2.7 Demand – Supply Gap in Stock

In 2014-15, the actual stock from supply is 98. Assuming outturn levels will grow at the same growth rate as observed during the last five years (*i.e.*, 6.56 per cent during 2010-11 to 2014-15), the estimated demand-supply scenario in 2025 will be:

Supply: 186

Demand: 364

Gap: 177 (51.00 % of the demand)

4.2.8 Flow Projections

The overall annual flow required in all sub-sectors of Agricultural engineering is given in Annexure-4.4 and summary in Fig. 4.3.

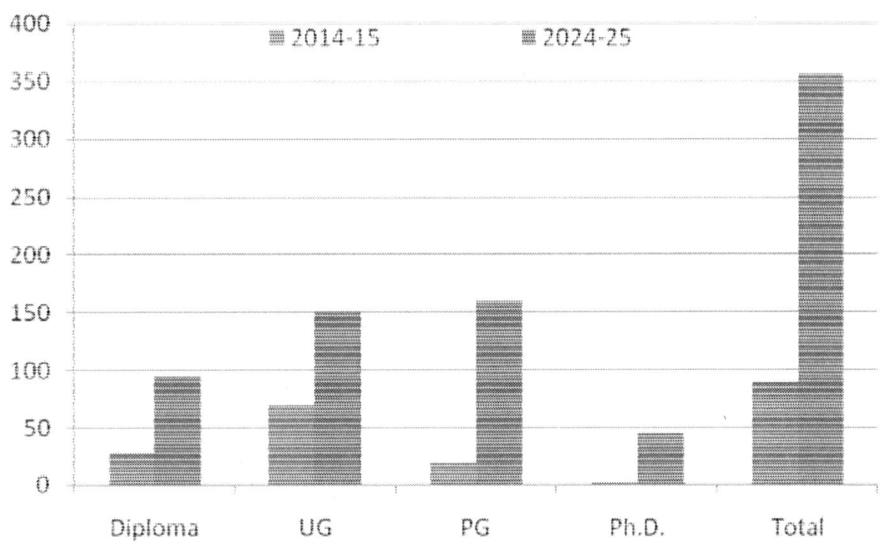

Fig. 4.3: Current (2014-15) and Projected Out turn Required (2024-25) in Agricultural Engineering

The projections suggest that by 2024-25, the annual out turn from the Agricultural Engineering education system should be about 120 diploma, 156 under-graduates 162 post-graduates and 45 doctorates, rising from 59 diploma, 74 under-graduates 23 post-graduates and 1 doctorate in 2014-15.

4.4 Human Resource Assessment in Agri-Biotechnology Sector

For about 10,000 years, farmers have been improving wild plants and animals through the selection and breeding of desirable characteristics. This breeding has resulted in the domesticated plants and animals that are commonly used in crop and livestock agriculture. In the twentieth century, breeding became more sophisticated, as the traits that breeders select for include increased yield, disease and pest resistance, drought resistance and enhanced flavour. Traits are passed from one generation to the next through genes, which are made of DNA. All living things—including the fruits, vegetables and meat that we eat—contain genes that tell cells how to function. Recently, scientists have learned enough to begin to identify and work with the genes (DNA) that are responsible for traits. Hence, agricultural biotechnology has come into existence for the collection of scientific techniques used to improve plants, animals and micro-organisms. Based on an understanding of DNA, scientists have developed solutions to increase agricultural productivity. Starting from the ability to identify genes that may confer advantages on certain crops, and the ability to work with such characteristics very precisely, biotechnology enhances breeders' ability to make improvements in crops and livestock. Biotechnology enables improvements that are not possible with traditional crossing of related species alone. A safe and sufficient food supply, grown in an environmentally responsible fashion, is essential for humanity. Like any technology, agricultural biotechnology will have economic and social impacts. Since their introduction, crops improved using biotechnology have been used safely, with benefits such as the reduction of pesticide use. Agricultural biotechnology is only one factor among many, influencing the health and welfare of farmers and other citizens in the developing world. As biotechnology continues to evolve, factual and open public discourse is vital to define the role, it should play in society. The Indian Biotechnology industry has an interesting evolution that has spanned over three decades. The milestone of the evolution of biotech industry in India began in 1978, in Bangalore, when the country's first biotech company, Biocon was established.

4.4.2 Education in Agri-Biotechnology

The largest challenges in Agri-biotechnology right now are public perception about biotechnology and GMO's. It is imperative that the impediments now obstructing innovations in these critical areas be examined, and those that cannot be justified must be removed. Where as, every tool in the box is now important. Biotechnology is not a threat. In order to supply the food requirement of the ever increasing population biotechnology is one of the better options to feed the hunger with available cultivable land. Hence, the trained human resource is the need of the hour. In Karnataka, education in Agriculture Biotechnology was started initially by offering two year post-graduation degree at College of Agriculture, UAS, Bengaluru and Dharwad during 1990s. Recently, the separate college of Agriculture Biotechnology was established by UAS, Bengaluru at Hassan campus and offering the four year B.Sc. (Biotech.) degree programme. Presently, it is only Agriculture Biotechnology College in the state. In future, few more colleges for Agriculture Biotechnology may be needed to meet the demand skilled human resource in the Agriculture Biotechnology sector.

4.4.1 Supply of Graduates

The out turn of graduates in UG and above in Agriculture Biotechnology are 91 (2014-15), among them 46 are under-graduate, 36 are post-graduate and 9 are doctoral degree graduates. Looking into past five years data of intake and out turn of graduates there is a increasing trend was observed 43 and 37 in 2010-11 to 59 and 46 in 2014-15 with CAGR 7.32 % and 5.78 %, respectively. Similarly, in master's degree, 31 and 30 in 2010-11 to 42 and 36 in 2014-15 with CAGR 7.62 and 5.26 % in intake and out turn, respectively. Whereas, 10 to 17 (2010-11 to 2014-15) with CAGR 12.26 % in intake and 8 to 9 (2010-11 to 2014-15) with CAGR 2.38 % in out turn of doctoral (Table 4.4.1).

Table 4.4.1: Intake and Out turn of Graduates in Agriculture Biotechnology Courses during

Year	Intake			Out turn			UG and above	
	UG	PG	Ph.D.	B.Sc.	M.Sc.	Ph.D.	Adm	Pass
2010-11	43	31	10	37	30	8	**84**	**75**
2011-12	52	37	10	37	25	10	**99**	**72**
2012-13	55	31	13	29	44	9	**99**	**82**
2013-14	56	42	11	42	29	10	**109**	**81**
2014-15	59	42	17	46	36	9	**118**	**91**
CAGR (%)	**7.32**	**7.62**	**12.26**	**5.78**	**5.26**	**2.38**	**8.07**	**5.17**

4.3.3 Areas of Placement for Agriculture Biotechnology Graduates

A number of well-qualified Agriculture Biotechnology graduates are required for the various sectors of the economy (public and private). Agriculture Biotechnology graduates were very much needed in tissue culture lab, veterinary pharmaceutical sector, academic sector, NGOs, in SAU's-Teaching, Research, Extension, seed companies, and other public institutions (Annexure-2).

4.3.4 Current Stock of Human Resources in Agriculture Biotechnology

There is no direct source of data on the number of active Agriculture Biotechnology human resources in the state. Hence, the same procedure was followed as that of section 5.1.4 (Agriculture sector). An effort was made to estimate the stock of Agriculture Biotechnology graduates from annual supply data from different colleges. The stock from supply side is estimated from the out turn of graduates in the past 5 years (details given separately in Annexure-4.1.4). The total stock of graduates and above in Agriculture Biotechnology, estimated from annual supplies is about 91 in 2014-15.

4.3.5 Demand Projections

Agriculture Biotechnology has emerged as one of the most important sector over the past few years. The demand for tissue cultured plantlets is growing rapidly. With its low cost skilled labour as well as scientific human resources (both of which are essential for tissue culture) has a natural advantage. Additional favourable factors are the wide range of plant biodiversity and favourable climate (which enables greenhouses with low energy consumption).

Till date about 94 commercial tissue culture production units have been recognized by the Department of Biotechnology (DBT), Govt. of India under the National Certification System for Tissue Culture Raised Plants (NCS-TCP). Among them, 13 Tissue Culture Plants (TCP) are there in Karnataka (**Annexure 4.5**). These TCP units are major employees of the Agriculture Biotechnology graduates in the state. These TCP units have been so far largely concentrating on exploiting the international markets and are facing a number of constraints such as short shelf life, stringent quality requirements and uncertainty of rejection of consignments, *etc.* Now-a-days the demand for Tissue culture Plants (TCP) is increasing at the rate of 12 per cent annual growth in the number of TCP plants likely to be demanded and also the number of units producing these plants. Three growth scenarios have been assumed - one with a high 12 per cent growth as projected by the Working Group and the other with a lower growth trajectory at 8 per cent per annum and average of two. On this basis, the demand per plants and the number of units have been projected as follows under the three growth scenarios (**Annexure 4.5**) and are summarized in Table 4.4.2.

Table 4.4.2: Requirement of Biotechnology Personnel in TCP Units

Year	Demand for biotechnologists			
	UG	PG	PhD	All UG & above
2014-15	10	13	3	25
2024-25	31	39	8	78

The demand for professionally trained graduate and above biotech professionals has been projected to grow from 25 in 2014-15 to 78 by 2024-25. Thus, the annual net addition to the stock is about 5 per annum. Detailed projections are given in Annexure 4.5.

Apart from the requirement of skilled human resource in TCP there is lot of demand for Agriculture Biotechnology graduates in seed and pharmaceutical sector Seed sector, in both public and private sectors, are working in harmony and showed considerable growth (4%) to meet the requirement huge quantity of seeds. Though public sector produces bulk of seed of varieties and self-pollinated crops, private sector is engaged in hybrid seeds and it too is playing a major role in producing large volumes of certified seeds of varieties for crops such as paddy, soybean, jute and hybrids of jowar, maize and bajra. Share of public and private sectors contribution is approximately 30 per cent & 70 per cent, respectively in terms of market turnover. Apart from the National Seeds Corporation, State Farms Corporation of India, Karnataka State Seed and Organic Certification Agency and other institutions in the public sector, there are about 125-135 private sector establishments/companies (http//:seednet.in) in the field of seed production and distribution. Information provided by the seed industry persons/experts indicated that the annual total requirement of Agriculture biotechnologists as agricultural scientists/experts in quality control is about 15-20 (3 Ph.D., 8 M.Sc. and 6 B.Sc.), Annexure 4.5.3. Similar requirement in agrochemical industries (Bio-pesticide and pharmaceutical industries) around 15 skilled graduates, Annexure 4.5.4.

Three scenarios have, therefore, been assumed, a high scenario with 12 per cent growth in employment and a low scenario with a growth of 8 per cent per annum and average of both. Based on these assumptions, the demand for biotechnologists in seed and agrochemical industries has been projected as shown in Annexure 4.5.4 and summary results are shown in Table 4.4.3.

Table 4.4.3: Requirement of Biotechnology Personnel in Seed and Agrochemical Sector

Year	Demand for Biotechnologists			
	UG	PG	PhD	All UG & above
Seed sector				
2014-15	10	6	1	16
2024-25	12	7	1	20
Agro-chemical sector				
2014-15	5	6	1	15
2024-25	16	19	2	47

The total staff sanctioned for teaching in Agriculture biotechnology discipline (6 from all SAUs in Karnataka) as on 2014-15 has been estimated at about 40, of which 42.50 per cent is found vacant (**Table-4.4.4**). Projection for the future years have been worked out assuming that the current vacancies will be filled to meet the requirement of the bio-tech personnel and the requirement would grow at the rate of 10 % for future 10 years.

The projected stock demand for State Agriculture University teaching for 2024-25 is 44, which would increase from 17 in 2014-15, Annexure 4.5.2 The annual requirement of teachers in SAUs is about 2 per annum.

Table 4.4.4: Overall Projections of Requirements Stocks of Agriculture Biotechnology Human Resources in SAU's

Year	UG	PG	Ph.D.	All UG & above
2014-15	1	2	14	17
2024-25	2	4	37	44

As per the information collected from experts and institutional survey, around 10 per cent of the total out turn of the agriculture biotechnologists are working in the other sectors. **The projected stock demand for other sector for 2024-25 is 39, which would increased from 20 in 2014-15 (Annexure 4.5.2) with the growth rate of 7 %.**

4.3.6 Summary of Demand Projections

The projections of stocks of Agriculture biotechnology human resources required to meet the needs of various segments of the economy and the consequential flows required from the educational system are summarized below:

4.2.6.1 Stock Projections

The overall stocks of Agriculture Biotechnology human resources required to meet the requirements of various segments of agricultural engineering have been added to give projections up to 2024-25. Three sets of projections have been worked out as explained in the segment-level projections - one a high projection, a low projection and an average. These projections are given in **Annexure 4.5**. The corresponding annual outputs required from the educational system to lead to these stocks are given in **Annexure-4.2.4.**

Considering the current supply and experts opinions on graduates aspirations and employee organizations capacity to absorb graduates, the average variant projection is considered desirable. The overall stock as per the average scenario is given in **Table-4.4.5.** The stock of graduates and above would grow from about 98 in 2014-15 to about 364 in 2024-25, *i.e.*, the average addition to the stock during the period 2015 to 2025 is about 40 per annum.

Table 4.4.5: Overall Projections of Requirements Stocks of Agriculture Biotechnology Human Resources

Year	UG	PG	Ph.D.	All UG & above
2014-15	45	38	22	108
2024-25	110	104	59	282

4.2.7 Demand – Supply Gap in Stock

In 2014-15, the actual stock from supply is 91. Assuming outturn levels will grow at the same growth rate as observed during the last five years (*i.e.*, 5.17 per cent during 2010-11 to 2014-15), the estimated demand –supply scenario in 2025 will be:

Supply: 152

Demand: 282

Gap: 130 (46.10% of the demand)

4.2.8 Flow Projections

The overall annual flow required in all sub-sectors of Agricultural Biotechnology is given in **Annexure-4.1.4** and summary in Fig-4.4.

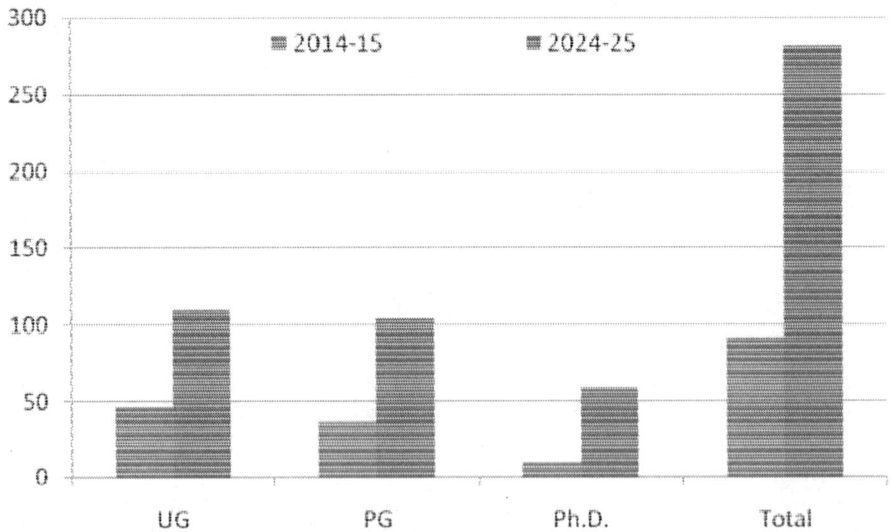

Fig. 4.4: Current (2014-15) and Projected Out turn (2024-25) in Agricultural Biotechnology

The projections suggest that by 2024-25, the annual out turn from the Agriculture, Agricultural Biotechnology education system should be about 110 under-graduates 104 post-graduates and 59 doctorates, rising from 45 under-graduates, 38 post-graduates and 22 doctorates in 2014-15.

4.5 Human Resource Assessment in Rural Home Science Sector

Home Science referred to as domestic science, home craft or domestic economy and was introduced in some schools and colleges during 1920 to 1940 of the British administration. Home Science education was initially designed and structured around the urban home- makers and her family and community. With the advent of Agricultural universities, the focus of Home science was shifted to the rural family with emphasis on intra-family resources, since rural families are composite and integrated units, which function as both consumption and production units.

Home Science consists of two words, that is, 'Home' and 'Science'. The word 'home' refers to the place of residence, where the family lives. The word 'science' refers to knowledge based on facts, principles and laws. By combining these two words, the meaning of Home Science can be derived as 'application of scientific knowledge in a systematic manner towards improving the quality of home and family life'. Hence, home science deals with providing education for family living; improving the efficiency of effectiveness of human and non-human resources used by families; conducting research for providing technologies as per changing needs of families and communities; helping the family and community to improve the quality of life. Rural Home Science gives knowledge for facing new challenges, to cope with knowledge explosion, technological advancements, new developments and growing needs of individuals for successful living in society. Home Science is one such subject that trains to face with confidence, the challenge of changing times for attaining satisfaction and harmonious living. It plays a key role in physical, physiological, psychological, social and economical development of individual, family and society.

4.4.2 Education in Rural Home Science

Home Science Education was initially designed and structured around the urban home-makers and her family and community. With the advent of Agricultural Universities, the focus of Home Science was shifted to the rural family with emphasis on intra-family resources, since rural families are composite and integrated units, which function as both consumption and production units. In Karnataka, College of Rural Home Science was started during 1974 with the primary mission is to impart quality education rural home science by delivering diversified subject-matter content in accordance with the spectrum of philosophy of individual and social institution for building, sensitizing and applying socially relevant multi-disciplinary knowledge and refining practices by developing scientific and technical skills for self-actualization and betterment of family and community in today's challenging world.

4.5.1 Supply of Graduates

The out turn of graduates in UG and above Rural Home Science are 83 (2014-15), among them 39 are under-graduate, 34 are post-graduate and 10 are doctoral degree graduates. Looking into past five years data of intake and out turn of graduates, increasing trend was observed 62 and 28 in 2010-11 to 91 and 39 in 2014-15 with CAGR 5.15 % and 5.84 %, respectively. Similarly, in master's degree, 26 and 22 in 2010-11 to 37 and 24 in 2014-15 with CAGR 13.74 and 4.56 % in intake and out turn, respectively. Whereas, 10 to 12 (2010-11 to 2014-15) with CAGR 5.82 % in intake and 8 to 10 (2010-11 to 2014-15) with CAGR 9.40 % in out turn of doctoral (Table 4.5.1).

Table 4.5.1: Intake and Out turn of Graduates in Rural Home Science Courses during 2010-11 to 2014-15

Year	Intake			Out turn			UG and above	
	UG	PG	Ph.D.	UG	PG	Ph.D.	Adm	Pass
2010-11	62	26	10	28	22	8	**98**	**58**
2011-12	90	19	9	33	16	7	**118**	**56**
2012-13	85	21	13	34	9	8	**119**	**51**
2013-14	69	34	11	30	21	11	**114**	**62**
2014-15	91	37	12	39	24	10	**140**	**83**
CAGR (%)	**5.15**	**13.74**	**5.82**	**5.84**	**4.56**	**9.40**	**7.02**	**8.53**

4.3.3 Areas of Placement for Rural Home Science Graduates

A number of well-qualified Rural Home Science graduates are required for the various sectors of the economy (public and private).

Rural Home Science graduates were very much needed in Developmental Departments, SAUs, Food Processing Industries, Hospitals, Self-Employment, NGOs, Financial Institutions and Insurance companies and Others sectors. Details were presented in the **Annexure-2**.

4.3.4 Current Stock of Human Resources in Rural Home Science

There is no direct source of data on the number of active Rural Home Science human resources in the state. Hence, the same procedure was followed as that of section **5.1.4** (Agriculture sector). An effort was made to estimate the stock of Rural Home Science graduates from annual supply data from different colleges. The stock from supply side is estimated from the out turn of graduates in the past 5 years (details given separately in 4.1.5). The total stock of graduates and above in Rural Home Science, estimated from annual supplies is about 83 in 2014-15.

4.3.5 Demand Projections

Rural Home Science has emerged as one of the most important sector over the past few years. The demand for Rural Home Science graduates in both public and private sector is increasing continuously. Rural Home Science graduates are recruited by many public sectors organizations and private companies in the state. The department of women and child development, Govt. of Karnataka having vacancies (27.30 %), at the state level the department is headed by Director, an IAS Officer, who is assisted by the Joint Directors in the four sections in the Directorate. Various level *viz.*, Administration, Integrated Child Development Services, Programme Development and Monitoring / Child Welfare, Stree Shakti / Women Welfare Programmes, *etc.* are the activities at state level and **district level** implementation of the district sector schemes is entrusted to Zilla Panchayats. At the Taluk level, the Child Development Project Officer is in-charge of implementation of the scheme and other programmes of the Department. Rural Home Science graduates are recruiting as the Assistant Child Development Officer and Project/Programme Officer. Presently, 40 graduates are working in the state departments, still lot of demand is there for the Rural Home Science graduates in various state departments. The detail of requirement s projected is given in the **Annexure-4.6.**

In agriculture universality, Rural Home Science graduates are required in the teaching, research and extension. There are about 22 vacancies (30.56 %) in all the six farm universities. Whereas, in private food processing industries, hospitals, NGO's and other sectors, around 50-55 graduates are required per annum. Detailed projections of requirements in the other sectors have been shown in **Annexure-4.6**.1 to 4.6.7.

Stock projections for the Rural Home Science graduates both in public and private have been shown in Annexure-4.6.1. The stock requirement of human capital would grow from 86 in 2014-15 to about 150 by 2024-25, *i.e.* on an average required 25 graduates more per annum.

Table 4.5.2: Staff positions in the developmental departments and SAU's as on March 31, 2015

Organisation	Sanctioned	Filled	Vacant	Vacancies %
SAUs	72	50	22	30.56
Department	55	40	15	27.27
Total	**127**	**90**	**37**	**29.13**
SAUs: Agriculture (UASB, D, R and S)				
Department: Department of women and child development, Govt. of Karnataka				

4.3.6 Summary of Demand Projections

The projections of stocks of Rural Home Science human resources required to meet the needs of various segments of the economy and the consequential flows required from the educational system are summarized below:

4.2.6.1 Stock Projections

The overall stocks Rural Home Science human resources required to meet the requirements of various segments of Rural Home Science have been added to give projections upto 2024-25. Three sets of projections have been worked out, as explained in the segment- level projections - one a high projection, a low projection and an average. These projections are given in **Annexure-4.6.**

Considering the current supply and experts opinions on graduates aspirations and employee organizations capacity to absorb graduates, the average variant projection is considered desirable. The overall stock as per the average scenario is given in **Table-4.5.3.** The stock of graduates and above would grow from about 83 in 2014-15 to about 184 in 2024-25, *i.e.*, the average addition to the stock during the period 2015 to 2025 is about 25 per annum.

Table 4.5.3: Overall Projections of Requirements Stocks of Rural Home Science Human Resources

Year	UG	PG	Ph.D.	All UG & above
2014-15	39	34	10	73
2024-25	100	88	50	240

4.2.7 Demand – Supply Gap in Stock

In 2014-15, the actual stock from supply is 73. Assuming outturn levels will grow at the same growth rate, as observed during the last five years (*i.e.*, 4.56 per cent during 2010-11 to 2014-15), the estimated demand -supply scenario in 2025 will be:

Supply: 128

Demand: 240

Gap: 112 (46.66 % of the demand)

4.2.8 Flow Projections

The overall annual flow required in all sub-sectors of Rural Home Science is given in **Annexure-4.1.5** and summary in Fig. 4.5.

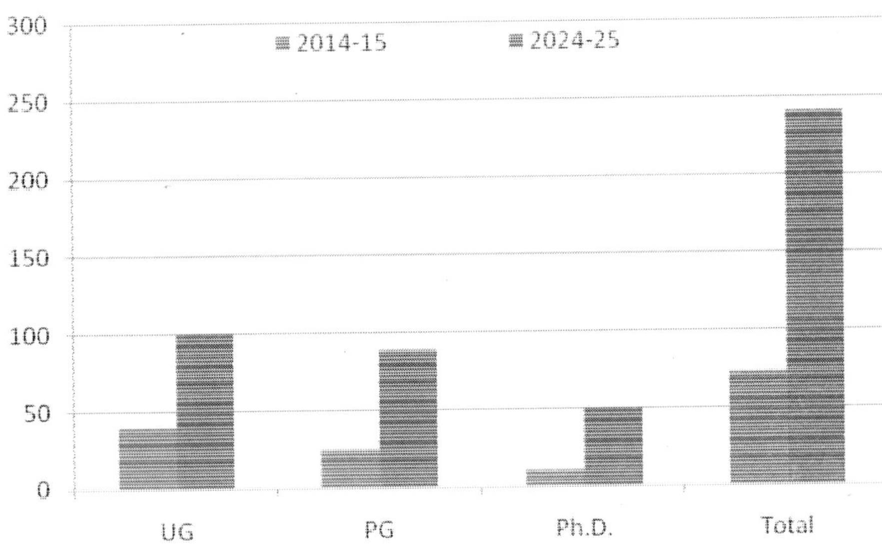

Fig. 4.5: Current (2014-15) and Projected Out turn (2024-25) in Rural Home Science

The projections suggest that by 2024-25, the annual out turn from the Rural Home Science education system should be about 100 under-graduates 88 post-graduates and 50 doctorates, rising from 39 under-graduates, 24 post-graduates and 10 doctorates in 2014-15.

4.6 Human Resource Assessment in Food Science and Technology

Food technology is a branch of food science that deals with the production processes that make foods. India is one of the most important food producers in the world. It is the leading producer of milk, pulses, sugarcane and tea and the second largest producer of wheat, rice, fruits and vegetables. Although food processing is one of the biggest industries in India ranking fifth in terms of production, consumption, export and expected growth, it accounts for only 1.7 % of world trade in this sector. The industry is relatively new, fragmented and unorganized at present. However, the changing lifestyles, food habits, urbanization and organized food retail have been driving the growth of processed food sector and poised to become one of the highest yielding sectors in India. In this fast growing era, processed food is of much importance, as it helps the new generation to tackle their life situations. Technical expertise in this sector can bring a practical situation that initiates possibilities in processing, packaging and storing of food without deteriorating its quality.

4.6.2 Education in Food Science & Technology

To meet the requirements for well-qualified and trained manpower for this fast growing industry, a need was felt for starting of the B. Tech. (Food Science and Technology) degree program. This degree program was started in the year 2007–08 and offers various courses in specialized areas of Food and Nutritional Science, Food Processing Technology, Food and Industrial Microbiology, Food Engineering and Food Trade and Business Management in addition to basic and advance courses in related areas. Importance is given to both for theoretical as well as practical knowledge. In Karnataka, post-graduation in Food Science & Technology degree programme was started in UAS, Bangalore and UAS, Raichur but intake capacity is very less, to meet the demand of well-qualified post-graduate persons, there is need to increase intake and also establish new colleges.

4.6.1 Supply of Graduates

The out turn of graduates in Food Science & Technology are 38 (2014-15). Looking into past five years data of intake and out turn of graduates, increasing trend was observed 49 and 31 in 2010-11 to 75 and 38 in 2014-15 with CAGR 12.01 and 6.63%, respectively (Table 4.6.1).

Table 4.6.1: Intake and Out turn of Graduates in Food Science & Technology Courses during 2010-11 to 2014-15

Year	Intake	Out turn
2010-11	49	31
2011-12	55	34
2012-13	71	45
2013-14	73	43
2014-15	75	38
CAGR (%)	**12.01**	**6.63**

4.3.3 Areas of Placement for Food Science & Technology Graduates

A number of qualified Food Science & Technology graduates are required for the various sectors of the economy (public and private). Food Science & Technology graduates were very much needed in Food Processing Industries as Food Technologist, Technical Specialist/ Manager, Process/Production Manager, QA Officer/Manager, Technical Sales Manager, Product Development Technologist, Food Scientist and Food Process Design Engineer; NGOs and others private sectors and for teaching in SAUs.

4.3.4 Current Stock of Human Resources in Food Science & Technology

There is no direct source of data on the number of active Food Science & Technology human resources in the state. Hence, the same procedure was followed as that of Agriculture sector. An effort was made to estimate the stock of Food Science & Technology graduates from annual supply data from college. The stock from supply side is estimated from the out turn of graduates in the past 5 years (details given separately in Annexure-4.1.6). The total stock of graduates and above in Food Science & Technology, estimated from annual supplies is about 75 in 2014-15.

4.3.5 Demand Projections

Food Science & Technology has emerged as one of the most important sector over the past few years. The demand for Food Science & Technology graduates in both public and private sector is increasing continuously. Food Science & Technology graduates employed mainly in private companies in the state. In SAU's skilled, Food Science & Technology personals are required for teaching. There are about 15 vacancies (42.86 %) in all the six farm universities **(Table 4.6.2)**. Whereas, in private food processing industries, NGO's and other sectors around 40 graduates are required per annum. Detailed projections of requirements in the other sectors have been shown in **Annexure-4.7.**

Stock projections for the Food Science & Technology graduates both in public and private have been shown in Annexure- 4.7.1 to 4.7.5. The stock requirement of human capital would grow from 66 in 2014-15 to 156 by 2024-25, *i.e.* on an average required 25 graduates more per annum.

Table 4.6.2: Staff positions in the SAU's as on March 31, 2015

Organisation	Sanctioned	Filled	Vacant	Vacancies %
SAUs	35	20	15	42.86
SAUs: Agriculture (UASB, D, R and S)				

4.3.6 Summary of Demand Projections

The projections of stocks of Food Science & Technology human resources required to meet the needs of various segments of the economy and the consequential flows required from the educational system are summarized below:

5.2.6.1 Stock Projections

The overall stocks Food Science & Technology human resources required to meet the requirements of various segments of Food Science & Technology, have been added to give projections upto 2024-25. Three sets of projections have been worked out as explained in the segment-level projections – one a high projection, a low projection and an average. These projections are given in **Annexure-4.7.**

Considering the current supply and experts opinions on graduates aspirations and employee organizations capacity to absorb graduates, the average variant projection is considered desirable. The overall stock as per the average scenario is given in **Table-4.6.3.** The stock of graduates and above would grow from about in 2014-15 to about 184 in 2024-25, *i.e.*, the average addition to the stock during the period 2015 to 2025 is about 25 per annum.

Table 4.6.3: Overall Projections of Requirements Stocks of Food Science and Technology

Year	UG	PG	Ph.D.	All UG & above
2014-15	75	0	0	75
2024-25	314	69	32	416

4.2.7 Demand – Supply Gap in Stock

In 2014-15, the actual stock from supply is 75. Assuming out turn levels will grow at the same growth rate as observed during the last five years (*i.e.*, 12.01 per cent during 2010-11 to 2014-15), the estimated demand-supply scenario in 2025 will be:

Supply: 233

Demand: 399

Gap: 166 (41.60 % of the demand)

4.2.8 Flow Projections

The overall annual flow required in all sub-sectors of Food Science & Technology is given in **Annexure-4.1.6** and summary in Fig. 4.6.

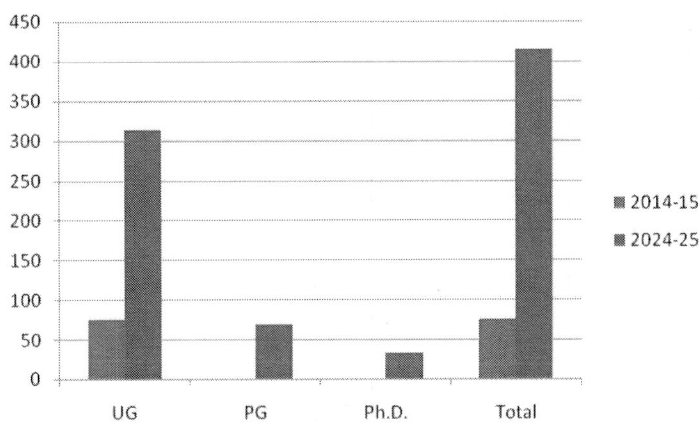

Fig. 4.6: Current (2014-15) and Projected Out turn (2024-25) in Food Science & Technology

The projections suggest that by 2024-25, the annual out turn from the Food Science & Technology education system should be about 399 rising from 75 in 2014-15.

4.9 Conclusions

The stock of graduates and above in agriculture sciences is estimated to increase from 1718 in 2014-15 to 3931 by 2024-25. Presently, the current annual increment of supply is 211. However, the demand for Agriculture graduates will increase to about 483 by 2024-25, *i.e.*, additional graduates (335 UG; 123 PG and 34 Ph.D.). To meet the existing demand Agriculture Marketing and co-operation, Agri. Engineering, Fisheries are needs opening up of additional colleges and increase of intake capacity in existing colleges. Issues related to course curriculum, teaching methodologies, development of soft skills and quality of teachers for the maintenance of quality in education. There is need of appropriate policy interventions for considering expansion of agricultural education in these sectors.

The annual supply for diploma holders is about 722 in 2014-15 and the requirement would grow to 2309 by 2024-25. Presently, private sectors are more absorbent of diploma level persons for field level works, in place of general graduates.

❑❑❑

Chapter 5

Human Resource Assessment in Horticulture Sector

5.1 Introduction

Horticulture is the branch of agriculture that deals with the art, science, technology, and business of growing plants. It includes the cultivation of medicinal plants, fruits, vegetables, nuts, seeds, herbs, sprouts, mushrooms, algae, flowers, seaweeds and non-food crops such as grass and ornamental trees and plants. It also includes plant conservation, landscape restoration, landscape and garden design, construction, and maintenance, and arboriculture. Inside agriculture, horticulture contrasts with extensive field farming as well as animal husbandry.

After the Green Revolution in mid-sixties, it became clear that horticulture, for which the Indian topography and agro climate are well-suited, is the best option. India has emerged as the largest producer of mango, banana and cashew and second largest producer of fruits & vegetables in the world. The most significant development that happened in the last decade is that horticulture has moved from rural confines to commercial production and this changing scenario has encouraged private sector investment in production system management. The last decade has seen technological infusion like micro-irrigation, precision farming, greenhouse cultivation, and improved post-harvest management impacting the development, but during the process various issues have emerged.

Horticulturists apply their knowledge, skills, and technologies used to grow intensively produced plants for human food and non-food uses and for personal or social needs. Their work involves plant propagation and cultivation with the aim of improving plant growth, yields, quality, nutritional value, and resistance to insects, diseases, and environmental stresses. They work as gardeners, growers, therapists, designers, and technical advisors in the food and non-food sectors of horticulture.

Horticulture sector has emerged as an important component of the economy of Karnataka state and has contributed more than one third share to the economy of agriculture and allied sectors. In many dry regions of the state, horticultural crops have evolved as an alternative crop to agricultural crops. Horticulture crops cover an area of 19.22 lakh hectare and the annual production is 162.57 lakh MTs (2014-15). The average productivity of horticultural crops in the state is 8.46 MTs per hectare. The annual value of horticultural products produced in the state is Rs. 35307 crore and constitutes 40.25% of the total income from entire agriculture sector. The share of horticultural produces in total GSDP of the state is 5.74%. In recent years, more emphasis is given for providing infrastructure, protected cultivation and Post-harvest Management. Introduction of advanced technology in cultivation of flower crops in Karnataka was started in 1990's. The progress made in this field attracted more farmers towards the cultivation of Roses, Carnation, Gerbera, Chrysanthemum, Orchids, Anthurium and other flower crops under greenhouse using advanced technology, which was lucrative in internal as well as international markets.

5.2 Education in Horticulture

Horticulture is one the fast growing sectors in Karnataka gaining greater importance due to its export demand, sustainable income, nutritional security and generation of employment. Realizing the importance of horticultural crops in Indian economy and of the state in particular, Under-graduate degree programme in Horticulture started in the year 1972-73 in Bangalore by UAS, Bangalore. Later one more horticulture college was established in Arabhavi by UAS, Dharwad during 1994. Realising the potential and importance of horticulture sector in the state Karnataka Govt., has approved the establishment of the University of Horticultural Sciences (UHS) at Bagalkot. Thus, the exclusive University of Horticultural Sciences came into being with effect from 22 November, 2008 at Bagalkot. This university has eight horticulture colleges and one Horticulture Engineering and Food Technology across the state. Later University of Agricultural and Horticultural Sciences, Shimogga was established during 2013. The College of Horticulture Mudigere (established in the year 1991) and COH, Hiriyur were transferred University of

Agricultural and Horticultural Sciences Shimoga from the year 2013. The horticulture education has been contributing in a positive way to leverage the development of producing necessary knowledge, technologies, varieties and skilled human resources required for horticulture industries, institutions and promoting self-farming by Horticulture graduates.

5.3 Supply of Graduates

The annual out turn of graduates in horticulture sciences are 404 (2014-15), among them 249 are under-graduate, 120 are post-graduate and 35 are doctoral degree graduates. Looking into past five years data of intake and out turn of graduates in horticulture science, the increasing trend was observed. The intake was 360 in 2010-11 and it is increased to 530 in 2014-15 with CAGR at 9.09 %, similarly in out turn was 180 in 2010-11 and it is increased to 249 in 2014-15 with CAGR at 13.63 % at under-graduate level. In master's degree, also increasing trend was observed in the intake 125 to 333 (2010-11 to 2014-15) with CAGR 26.42 % and out turn 75 to 120 (2010-11 to 2014-15) with CAGR 10.58 %. Similarly, in Doctoral degree intake was increased 31 to 66 (2010-11 to 2014-15) with CAGR 23.20 % and out turn 20 to 35 (2010-11 to 2014-15) with CAGR 15.00 % (Table 5.1).

Table 5.1: Intake and Out turn of Graduates in Horticulture Courses during 2010-11 to 2014-15

Year	Intake				Out turn				UG and above	
	Diploma	UG	PG	Ph.D.	Diploma	UG	PG	Ph.D.	In	Out
2010-11	45	360	125	31	22	180	75	20	**516**	**275**
2011-12	50	445	211	36	25	200	89	18	**692**	**307**
2012-13	73	470	441	53	31	300	90	21	**964**	**411**
2013-14	61	490	310	64	45	375	95	38	**864**	**508**
2014-15	61	530	333	66	50	249	120	35	**929**	**404**
CAGR (%)	**8.41**	**9.09**	**26.42**	**23.20**	**24.98**	**13.63**	**10.58**	**20.52**	**15.00**	**13.57**

5.4 Areas of Placement for Agriculture Graduates

Graduates with a degree in Horticulture enter a broad range of challenging and rewarding professional careers in both public and private sectors. In public sector Horticulture Department, and Universities-Education, Research and Extension are the major employers. In private, production of planting material, establishment of orchards, post-harvest storage, quality control, transport, cold storage, and processing of horticulture commodities, Seeds companies/

seed farms, Agro Chemical Industries, Financial institutions, Post-harvest-quality Control, NGOs and others.

5.5 Current Stock of Human Resources in Horticulture Science

There is no direct source of data on the number of active Horticulture science human resources in the state. As such, the current stock can only be estimated, as in the case of other agriculture related disciplines, indirectly by cumulating the out turn of graduates, post-graduates and doctorates in Horticulture sciences over past five years (as many as the active working life of a graduate, post-graduate or doctor) and adjust for attrition due to retirement, migration and such causes.

An effort was made to estimate the stock of Horticulture science graduates from annual supply data from two horticulture universities. The stock from supply side is estimated from the out turn of graduates in the past 5 years. The total stock of graduates and above in Horticulture sciences, estimated from annual supplies is about 404 in 2015.

5.6 Demand Projections

Horticulture is fast growing field of agriculture that deals with the cultivation and processing of plants, fruits, vegetables, flowers, aromatic and medicinal plants. Now-a-days, people attraction towards horticulture sector has increased because government focusing on horticultural sector due to its significant contribution in the economy as employment generators and industrial raw material provider. Liberalization, privatization and globalization of world trade also generated the opportunities of employment for youths across the country. Present study mainly focuses on assessment of employment opportunities for horticultural professionals in both government and private sector. The candidate after completion of graduation or post-graduation is eligible for many governmental and private jobs in academics and research areas of horticultural sector. Thus, a horticultural graduate can get employment on various posts such as horticulturist, floriculturist, olericulturist, pomologist, technician, horticultural consultant, horticultural therapist, botanical garden worker, greenhouse manager, landscape architect, viticulturist, nursery manager and horticultural marketers in governmental and private sectors.

5.6.1 Development Departments

Karnataka is the first state in the entire country to have a separate Horticulture Department, and many other states, at later years, followed the example of Karnataka. Because of this, the state could achieve remarkable progress in many fronts of Horticulture, whether fruits or vegetables or flowers or plantation crops. The growth of Horticulture in the state has a thrilling historical background, which reveals several interesting facts, milestones of achievements and pride legacies. Although some horticulture crops were grown in the state since time immemorial, their cultivation on commercial scale started just two and a half century before. The first acknowledgeable credit for initiation of cultivation of horticulture crops in the state undoubtedly goes to Haider Ali and Tipu Sultan (1970). In August 1856, Lalbagh was made the Government Botanical Garden, becoming entirely a government establishment. The real developmental works in Lalbagh started from the year 1874, when John Cameron took charge of the garden. In 1963, consequent to the formation of the separate Department of Horticulture, he was elevated to the post of the Director of Horticulture. Several schemes, which were earlier handled by the Agriculture Department were transferred to this newly created Department. In 1965, the reorganization of the Department took place and several posts were created in order to carry over the task of horticulture development in the state. In 1956, as a result of reorganization of the state, the horticultural activities were extended to all the 19 districts. Thus, he was responsible for elevating the minor Department of Horticulture to a major Department.

To suit horticulture to the rural areas and to the common man, the implementation of an unique pattern of Horticulture development *i.e*, "4-Limbed Model of Horticulture. By establishment of Horticulture Produce Co-operative Marketing Society and the Nursery men's Co-operative Society at Bangalore. Later, 357 farms and nurseries started all over the state which leads to development of progeny orchards, nursery centers and places of demonstration of new crops and technology to the farmers. Seed testing, soil testing and plant protection laboratories were started at Lalbagh. Several parks and gardens were laid out in different cities and towns of the state.

The area of Lalbagh Botanical Garden was expanded to 240 acres and planted with additional native and exotic species of plants during his period. Dr. Marigowda was a staunch advocator of Dry Land

Horticulture and the principals and practices of these technologies were demonstrated in most of the farms started by him. This inspired the farmers of the state to practice Dry Land Horticulture on vast dry and drought prone tracts of the state. Mixed cropping and inter-cropping got special fillip during his times. Thus, through multifarious achievements and feats, the state of Karnataka became the "Horticultural State of India", and Dr. Marigowda's name became immortal in the annals of Horticulture development in Karnataka.

In order to achieve the target of the department of horticulture, there is lot of demand for skilled horticulture graduates in the state. The data related to the staff positions in the horticulture department have been collected (sanctioned filled and vacant positions) and estimates have been made on that basis. The estimate showed that there is a 40.59 % vacancy in the state. The number of horticulture officers in the state has been estimated at 1035 (Table 5.2).

Detailed projected stock demand in department of horticulture is shown in Annexure-4.11.1. The estimated stock requirement in 2014-15 is about 707. This would increase to 1152 by 2024-25. Thus, department of horticulture would need about 70 per annum during 2015 to 2025.

5.6.2 State Horticulture / Agriculture Universities

Different Agricultural/ Horticulture universities in the state employ horticultural post-graduates/ doctoral for distinct posts from concerned field of their specialization. However, many numbers of posts of associate professor, assistant professor and other teaching, research and extension are vacant. To meet the technical manpower required for the Horticulture Industry of the State, a three-tier system of education is drawn with emphasis on the; Imparting education towards development of quality human resource in different branches of horticulture. Further, advancement of learning and conducting of research in horticulture and other allied sciences, undertaking the extension education of such science and technologies, specially for the rural people of the state. Promoting partnership and linkages with national and international educational, research institutions and industries. The SAUs (including all teaching, research and extension) had a combined sanctioned strength of 893 faculties, of which 585 positions (34.45 %) were vacant as on March 31, 2015 details was given in (Table 5.2).

Detailed projected stock demand in farm universities is shown in Annexure-4.11.2. **The estimated stock requirement in 2014-15 is about 308. This would increase to 502 by 2024-25. Thus, SAU's including Teaching, Research and Education; needs about 20 persons per annum during 2015 to 2025.**

Table 5.2: Staff positions in the horticulture department and farm universities as on March 31, 2015

Organisation	Sanctioned	Filled	Vacant	Vacancies %
Universities	893	585	308	34.45
Department	1742	1035	707	40.59
Total	**2635**	**1620**	**1015**	**37.52**
Universities - Agriculture and Horticulture				
Department: Horticulture Department, Govt. of Karnataka				

5.6.3 Financial Institutions

The job opportunities are accessible for the students within the country after graduation/post-graduation. The nationalized banks, reserve bank, private banks put forward an opportunity for post-graduates in horticulture as agricultural officers, probationary officers, field officers and rural development officers by conducting exam through IBPS. Establishment survey showed that about 4 per cent of the total employment (or, 10% of the officers in banks, as officer strength is 40 per cent of the total employment) in banks comprised graduates and post- graduates in agriculture and allied sciences. About 89 per cent of them were graduates and 3 per cent post-graduate, doctorates forming a mere 0.2 per cent, the rest being diploma/certificate holders in agri-related fields. Employment of officers in public sector banks has been projected to grow at a rate of 2.5 per cent per annum.

Data on how many of the agricultural graduates employed by the banking institutions are horticulture qualified is not known as the banking institutions do not generally prescribe specific qualifications in horticulture at the time of recruitment, it may be assumed that the recruitment would be in proportion of the horticultural graduates, post-graduates to the total annual intake of graduates in all agricultural fields (which is about 5 per cent, now and may go upto 10% by 2025) as per research report (Rama Rao *et al.*, 2010).

Detailed stock projections for financial institutions are given in Annexure-4.11.3. The stock requirement of human capital would grow from 35 in 2014-15 to about 52 by 2024-25. Thus, the requirement is about 3 per annum.

5.6.4 Private Sector Employment

Horticultural graduates are employed in various private companies *viz.*, seed sector, fertilizer and pesticide industries, offering R&D, production and marketing jobs. Horticulture graduates have contemporary, challenging career in private sectors. The horticulture industry has spawned a number of supporting or service industries such as (1) **Research:** Many scientists are engaged in developing new and improved types of vegetables, fruits and ornamentals. Apart from these, horticultural scientists also devote considerable time to improve the aesthetics of ornamentals and the quality of products. (2) **Chemical industry:** The horticulture industry depends on large amounts and varieties of chemicals, including fertilizers, pesticides and growth hormones. Many companies are involved in producing chemicals that are used to enhance plant production and the quality of produce. (3) Production of essential oils. Many firms are involved in extraction of medicinal and aromatic plants essential oils, which is used in confectionary, cosmetic, medicine, beverage and food industries at large scale. (4) **Machinery:** The engineers design and produce tool and machinery for use in the production of horticultural plants. Machinery and implements are available for preparing, harvesting, storing and packaging. These aids enable large scale production of horticultural plants to be undertaken. (5) **Distribution:** Horticultural products are transported from the areas of production to marketing outlets. Horticulture produce and product require special handling on transportation due to their perishability to retain their quality for a long time. (6) **Computer in horticulture:** Computers are widely used in the horticultural field such as landscape designing, crop modelling and soil modelling equipment automation, public information, record keeping and databases. Expert systems, communication, sensor technology and internet services, *etc.* computer technology have significant applications in horticulture.

On the basis of the data in Annexure 4.1.2 and Tables-5.3, Tables-5.4 & 5.5, projections have been made of the horticultural human resources needed till 2024-25 in seed & agrochemical sectors and horticultural production & processing sector.

It may be observed from these projections that the seed & agrochemical sectors, horticulture production and processing industries are expected to employ 122, 57 and 104 by 2024-25, respectively. The stock requirement of human capital would grow

about 10, 15, 20 per annum during the period 2024-25 (seed & agrochemical sectors, horticultural production, and processing sector), respectively.

Table 5.3: Projected Horticulture Human resource requirement in seed & agrochemical sectors till 2024-25

Year	Diploma	UG	PG	Ph.D.	Total
2014-15	60	30	41	4	75
2024-25	98	49	67	6	122

Table 5.4: Projected Horticulture Human resource requirement in horticultural production sector till 2024-25

Year	Diploma	UG	PG	Ph.D.	Total
2014-15	40	19	14	2	35
2024-25	65	31	23	3	57

Table 5.5: Projected Horticulture Human resource requirement in processing sector till 2024-25

Year	Diploma	UG	PG	Ph.D.	Total
2014-15	30	18	22	2	40
2024-25	49	47	57	5	104

5.6.5 NGOs

A number of NGOs are undertaking extension and promotional work in agriculture in general and horticulture in particular. There are also a number of agro-service centres providing professional services and inputs like seeds, fertilizers and equipment to farmers. In recent years, the scheme of agri-clinics and agri-businesses to provide self-employment to agricultural graduates has also led to rapid growth of such clinics. All these developments have an impact on horticulture also NGO's engaged in extension of horticulture activities such as mushroom cultivation, apiary, vegetable production and marketing, tissue culture, seed processing and marketing, nursery, landscaping and nursery, floriculture and cultivation of medicinal plants are employed horticulture graduates. Thus, on an average 10 units have come up per year, each generally with one horticulture graduate.

5.6.6 Self-Employment and Entrepreneurship Development

After graduation and post-graduation, a horticultural graduate can start the their own horticulture clinic, which are envisaged to provide expert advice and services to farmers on technology, cropping practices, protection from pests and diseases, market trends, prices of fruits and vegetable crops and also clinical services for animal health, which would enhance productivity of crops/animals and increased income to farmers. A graduate can also start the business of nursery raising of fruit plants, ornamental plants, aromatic and medicinal plants and bonsai by the plant material using various techniques such as sexual propagation (seed) and asexual propagation like grafting, budding, layering and tissue culture. Horticulture provides immense potential of entrepreneurship development in nursery management, landscaping, processing, marketing and protected cultivation of higher value horticultural crops. A horticultural graduate can start their own retailing business of fruits, vegetables and flowers using mobile based agri-retailing. As per establishment survey and discussion with experts around generally one to two horticulture graduate are becoming Self-Entrepreneur, still lot of demand is there to increase up to 10 per annum.

5.6.7 Other Establishments

There are a number of other societies, boards and institutions also recruit the horticulture graduates such as Commodity boards, management institutions, business agencies and foundations at national and international level. **Detailed projections of requirements in the other sector have been shown in Annexure-4.11.9. The stock requirement of human capital would grow from 25 in 2014-15 to about 54 by 2024-25 *i.e.* 2 per annum.**

5.7 Summary of Demand Projections

The projections of stocks of horticulture human resources required to meet the needs of various segments of the economy and the consequential flows required from the educational system are summarized below:

5.7.1 Stock Projections

The overall stocks of horticulture human resources required to meet the requirements of various segments of horticulture have been added to give projections upto 2024-25. Three sets of projections have been worked out as explained in the segment- level projections – one a high projection, a low projection and an average. These projections are given in Annexure-4.11.

Considering the current supply and experts opinions on graduates aspirations and employs organizations capacity to absorb graduates, the average variant projection is considered desirable. The overall stock as per the average scenario is given in **Table-5.6.** The stock of graduates and above would grow from about 415 in 2014-15 to about 681 in 2024-25, *i.e.*, the average addition to the stock during the period 2015 to 2025 is about 35 per annum.

Table 5.6: Overall Projections of Horticulture Human Resources

Year	Diploma	UG	PG	Ph.D.	All UG & above
2014-15	50	249	120	35	**404**
2024-25	179	893	328	226	**1443**

5.7.2 Demand – Supply Gap in Stock

In 2014-15, the actual stock from supply is 404. Assuming out turn levels will grow at the same growth rate as observed during the last five years (*i.e.*, 13.54 per cent during 2010-11 to 2014-15), the estimated demand–supply scenario in 2025 will be:

Supply: 1443

Demand: 2118

Gap: 674 (31.82 % of the demand)

5.7.3 Flow Projections

The overall annual flow required in all sub-sectors of horticulture is given in Annexure-4.1.9 and summary in Fig. 5.1.

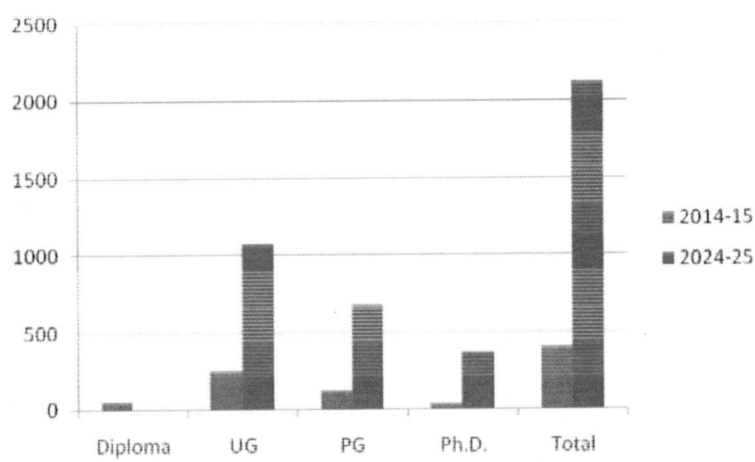

Fig. 5.1: Current (2014-15) & Projected Out turn Required (2024-25) in Horticulture

The projections suggest that by 2024-25, the annual out turn from the horticulture education system should be about 179 diploma, 893 under-graduates 328 post-graduates and 226 doctorates, rising from 50 diploma, 249 under-graduates 120 post-graduates and 35 doctorates in 2014-15.

5.8 Conclusion

Current annual out turn of about 404 graduates is less than the requirement of 2118 by 2024-25. There is need to expand horticulture education to produce additionally 674 graduates (180 UG; 349 PG and 141 Ph.D.) per year by 2024-25. This needs opening up of some more additional colleges across the state.

The diploma holders are needed in private sector (micro-level organizations) because diploma holders would be preferred in place of general graduates now, being employed by many employers, especially for some field based rural jobs. The annual supply of diploma holders is about 50 in 2014-15 and the requirement would grow to 260 by 2024-25.

□□□

Chapter 6

Human Resources Assessment in Forestry Sector

6.1 Introduction

Forest has been defined as 'A large area covered chiefly with trees and undergrowth'. The Webster dictionary has also defined 'forest' as a dense growth of trees and underbrush covering a large tract. Karnataka is endowed with some of the most magnificent forests in the country. From the majestic evergreen forests of the Western Ghats to the scrub jungles of the plains, a wide variety of habitats exist with very typical flora and fauna, some of them endemic to the region. The Western Ghats, which covers about 60% of forest area of Karnataka, is recognized as one of the 18-mega bio-diversity hotspots of the world. The total recorded forest area of the State is 38,284.30 sq. km., constituting 20% of the geographical area. The Western Ghats forests are rich in fauna and floral diversity and are very valuable as both genetic and natural resources. Many plant and animal species are endemic to forests in the state. Several economically important species such as Sandalwood (*Santalam album),* Rosewood (*Dalbergia latifolia*), Teak (*Tectona grandis*), Honne (*Pterocarpus marsupium*), White Cedar (*Dysoxylon malabaricum*) and many non-timber and other medicinal plant species grow naturally in the forests of Karnataka. The state has a protected area network of 5 national parks (2431.300sq.km.), and 21 wildlife sanctuaries (3887.827 sq. km), covering nearly 16% of its forest area. The forests of Karnataka are rich in biodiversity and support 25% of the elephant population and 10% of the tiger population of the country, in addition to numerous species of mammals, birds, butterflies, amphibians, reptiles, flowering plants, *etc.*

The conservation of forest wealth has become a major challenge in the whole of the country, including the state of Karnataka, because of ever-increasing pressures on already declining forest resources. As a result of burgeoning population and consequent landlessness, the pressure on the available land resources has been increasing day by day leading to encroachment of forestland, which has become a major threat to the conservation of forest and its resources. In many places, it has led to fragmentation of forest areas, honeycombing of forest areas, loss of corridor for movement of wild animals, *etc.* Also in Karnataka, there are valuable patches of private forests that require protection. Smuggling of timber and poaching of wild animals have now emerged as the most serious threats to the forest resources of the state. It has become difficult to manage these threats with the existing manpower, infrastructure and financial resources.

6.2 Education in Forestry

Realizing the importance of skilled human resources in this sector, programmes have been initiated in various aspects of forestry. The Universities initiated organized forestry education only in 1987, in India. In Karnataka college of Forestry, Sirsi is one of the premier institutes of forestry education in the country and a constituent college of the University of Agricultural Sciences, Dharwad. The college is situated in one of the hotspots of biodiversity in the world - the Western Ghats. College of Forestry has an immense role to play in identifying, documenting and conserving biological diversity, which is unique to this part of the country with a mission to become a premier institute in developing human resource in the field of forestry who will be responsible and socially committed to the real stakeholders. Since its inception at College of forestry, Sirsi, it is committed in preparing the students at under-graduate and post-graduate level to be professionals and teachers, researchers and planners managers and policy-makers who will contribute throughout their careers to viable and sustainable management of state forests and natural resources. Later Forestry College, Ponnampet was established in the year 1995; was originally functioning at GKVK, Bangalore upto the year 1994 and then shifted to the present campus at Ponnampet coming under UAHS, Shimoga. Presently, in under-graduate degree programmes and Post-graduate degree programmes in various disciples of forestry are being offered in these two colleges.

6.3 Supply of Graduates

The out turn of graduates in forestry are 99 (2014-15), among them 60 are under-graduate, 35 are post-graduate and 4 are doctoral degree graduates. Looking into past five years data of intake and out turn of graduates in forestry, the decreasing trend in intake 115 in 2010-11 to 110 in 2014-15 with CAGR at 0.80 %, whereas increasing trend in out turn 30 in 2010-11 to 60 in 2014-15 at under-graduate level with CAGR at 22.25 %. In master's degree, increasing trend existed in the intake and out turn 30 and 40 in 2010-11 to 23 and 35 in 2014-15 with CAGR 9.49% and 6.17%, respectively. Similarly, increasing trend observed in the intake and out turn of doctoral 5 and 6 in 2010-11 to 2 and 4 in 2014-15 with CAGR 5.62 and 37.41%, respectively (Table 6.1).

Table 6.1: Intake and Out turn of Graduates in Forestry Courses during 2010-11 to 2014-15

	Intake			Outturn			UG and above	
Year	UG	PG	Ph.D.	UG	PG	Ph.D.	In	Out
2010-11	115	30	5	30	23	2	**150**	**55**
2011-12	115	28	5	44	28	1	**148**	**73**
2012-13	163	27	5	106	22	5	**195**	**133**
2013-14	116	39	6	82	22	6	**161**	**110**
2014-15	110	40	6	60	35	4	**156**	**99**
CAGR (%)	**0.80**	**9.49**	**5.62**	**22.25**	**6.17**	**37.41**	**1.64**	**17.18**

6.4 Areas of Placement for Forestry Graduates

Number of well-qualified forestry graduates is required in the Forestry Department, Govt. of Karnataka apart from Agricultural Universities for teaching, research and extension, Administrative posts, Agro Processing- Timber & Wood Industry, NGOs and Other sectors.

6.5 Current Stock of Human Resources in Forestry

There is no direct source of data on the number of active forestry graduates human resources in the state. Hence, the same procedure was followed as that of section **5.1.4** (Agriculture sector).

An effort was made to estimate the stock of forestry graduates from annual supply data from the two colleges. The stock from supply side is estimated from the out turn of graduates in the past 5 years (details given separately in Annexure-4.1.8). The total stock of graduates and above in forestry, estimated from annual supplies is about 99 in 2014-15.

6.6 Demand Projections

Forests and people are connected and have been since ancient times. We have always had a special relationship based on survival. It was a delicate chain of existence that we once treated with respect and appreciation. The National Forest Policy, 1988 recommends coverage of at least 33% of the geographical area under forest & tree cover to maintain the ecological balance of any region. To achieve this, it is not only necessary to bring new areas under afforestation activities but also it is necessary to protect and consolidate the existing forest areas. There are several projects in vogue for the purpose of afforestation. The efforts of protecting, developing and consolidating natural forests are also going on. However, the infrastructure and resources are not in proportion to the problems faced and hence, the results are not as expected. In order to maintain the forest cover as expected level, it is necessary to recruit more and more personnel trained in forestry sector.

The infrastructure and staffing available with forest department is inadequate. Increased biotic pressure and use of vehicles, modern communication equipments and weapons by forest offenders have made the task of forest protection indeed a difficult one. It is necessary that the infrastructure available to the staff should be of modern nature to meet the challenges in the field.

Ever increasing demand for forest produce like firewood, grasses, bamboos and minor forest produce (NTFP) from already shrinking forest areas, has led to degradation of forest stocks in quality and quantity. In several areas the rootstock available, if protected, can regenerate into well-stocked biodiversity areas. In other areas, valuable species have disappeared giving way to blanks and invasive species. Here apart from protection, augmentation of the stock with valuable species by the way of planting is necessary. Thus, by protecting the existing rootstock, production of quality planting material and planting it to augment the stock along with natural regeneration can improve the existing stock to a desirable level.

The forest areas notified or otherwise are widely scattered and are spread over most difficult terrains, their long boundaries make it difficult to maintain them and keep track of invading neighbourer resulting in encroachment of forestlands. For several decades, survey and demarcation of forestlands have not been taken up. It is of utmost importance to know the extent and limits of the asset, we manage before doing anything.

Hence, survey and demarcation of forest areas and consolidation of the same by creating barriers in the form of fences, boundary trenches, etc., is a necessity. Karnataka is bestowed with very rich fauna & flora because of variety of habitats in the form of Western ghats, Eastern ghats, Eastern plains, and Coastal areas. Several of the species are endemic to the State. Increasing population and faster developmental activities as a part of globalization process have put immense pressure on the flora and fauna. Effective conservation measures taken up have resulted in protection and increase of wildlife populations in certain pockets. Invading trend of human activities combined with spilling of wildlife populations, out of their habitat has resulted in severe man wildlife conflicts. To achieve co-existence of human and wildlife, it is necessary to improve the wildlife habitat to contain the wildlife in the forest areas. It is also necessary to provide protective barriers not to allow the spilling of increased wildlife populations beyond the forest areas. Hence, there is an immense potential demand for forestry graduates both in public and private sector.

In the Forest Department, Govt. of Karnataka, the skilled human resources are required in various posts *viz.*, Forest Officer, Deputy Range Forest Officer (DRFO), Range Forest Officer (RFO), Asst. Conservator of Forests (ACF), Deputy Conservator of Forests (DCF) and Chief Conservator of Foresters (CCF), *etc.* The Department is headed by the Principal Chief Conservator of Forests (Head of Forest Force). Each of the Revenue Districts has a Social Forestry Division, headed by a Deputy Conservator of Forests, attached to Zilla Panchayat administratively. Social Forestry Divisions take up afforestation outside the notified forest areas and promote farm forestry and agro-forestry. The Wildlife protected areas are organized into Wildlife Sanctuaries, National Parks, Conservation Reserves and Community Reserves. Wildlife wing of the Department is headed by Principal Chief Conservator of Forests (Wildlife) & Chief Wildlife Warden. In addition to the above, the department has required skilled human resource for, Research, Training, Evaluation, Vigilance, ICT cell, and Mobile Squads, as functional units, *etc.*

The data related to the staff positions in the Forest department have been collected (sanctioned, filled and vacant positions) and estimates have been made on that basis. The estimate showed that there is a 34.28 % vacancy in the state. The number of forest officers in the state has been estimated at 2402 (Table 6.2).

Detailed projected stock demand in the Forest department is shown in Annexure-4.9. The estimated stock requirement in 2014-15 is about 1269. This would increase to 2067 by 2024-25. Thus, department of Forestry would need about 70 per annum during 2015 to 2025.

Table 6.2: Staff positions in the Forest department and SAU's as on March 31, 2015

Organisation	Sanctioned	Filled	Vacant	Vacancies %
SAUs	99	53	46	46.84
Department	3702	2403	1269	34.28
Total	3801	2456	1315	40.56
SAUs- Agriculture (UASB, D, R and S)				
Department: Department of Forestry, Govt. of Karnataka				

6.7 Summary of Demand Projections

The projections of stocks of forestry human resources required to the meet the needs of various segments of the economy and the consequential flows required from the educational system are summarized below:

6.8 Stock Projections

The overall stocks of forestry human resources required to meet the requirements of various segments of forestry have been added to give projections up to 2024-25. Three sets of projections have been worked out as explained in the segment- level projections – one a high projection, a low projection and an average. These projections are given in Annexure-4.9. The corresponding annual outputs required from the educational system to lead to these stocks are given in **Annexure-4.9.10.**

Considering the current supply and experts opinions on graduates aspirations and employee organizations capacity to absorb graduates, the average variant projection is considered desirable. The overall stock as per the average scenario is given in **Table-6.3.** The stock of graduates and above would grow from about 415 in 2014-15 to about 681 in 2024-25, *i.e.*, the average addition to the stock during the period 2015 to 2025 is about 35 per annum.

Table 6.3: Overall Projections of Requirements Stocks of Forestry Human Resources

Year	UG	PG	Ph.D.	All UG & above
2014-15	60	35	4	99
2024-25	955	451	56	1463

In 2014-15, the actual stock from supply is 99. Assuming out turn levels will grow at the same growth rate as observed during the last five years (*i.e.*, 17.06 per cent during 2010-11 to 2014-15), the estimated demand–supply scenario in 2025 will be:

Supply: 497

Demand: 1308

Gap: 811 (62.00 % of the demand)

6.10 Flow Projections

The overall annual flow required in all sub-sectors of forestry is given in Annexure- 4.1.8 and summary in Fig. 6.1.

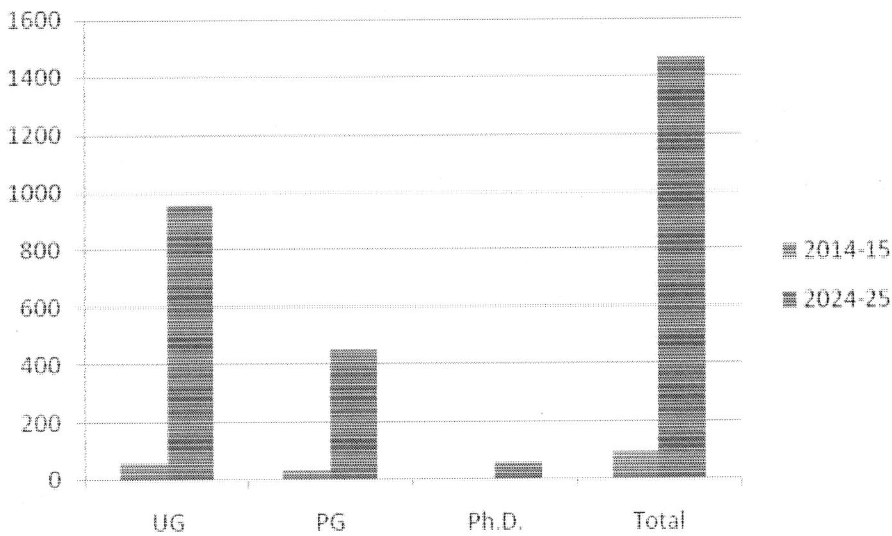

Fig. 6.1: Current (2014-15) and Projected Out turn required (2024-25) in Forestry

The projections suggest that by 2024-25, the annual outturn from the forestry education system should be about 955 under-graduates 451 post-graduates and 56 doctorates, rising from 60 under-graduates 35 post-graduates and 4 doctorates in 2014-15.

6.11 Conclusion

Current annual out turn of about 99 graduates is very less than the requirement of 1463 by 2024-25. There is need to expand forestry education to produce additionally 1137 graduates (72 UG; 37 PG and 4 Ph.D. per year) by 2024-25. This needs increase intake capacity of existing colleges in the state. In fact, the current intake capacity is about 110 and students are not readily opting for this course. Unfortunately, graduates in this discipline have no private sector to absorb, unlike in other subjects like agriculture, horticulture, fishery, *etc.* Even in academic and research sector, the situation is no good except in ICAR and SAUs, where the positions are filled with forestry graduates. However, these are large number of vacancies in the Department of Forestry, which requires special trained in forestry science.

The rural institutions like Van Vigyan Kendras and Joint Forest Management activities at grassroot level and a host of government schemes aimed at forest and tribal areas, need forest trained persons at village level. This requirement can be met by diploma holders. Hence, it is suggested that to start diploma and certificate courses to fulfill the requirements.

□□□

Chapter 7

Human Resources Assessment in Sericulture Sector

7.1 Introduction

Many countries in the world both developed and developing are engaged in the production of silk and silk goods. During the past few decades, the silk-producing countries have witnessed dramatic changes in their structure of economic profiles. Some countries, which were in the forefront in silk-production, have ceased to produce enough output for their consumption, while some are importing to meet their domestic needs, and some others produce a marketable surplus for export. Some countries are struggling to produce quality silk and export. Karnataka is major silk-producing state in India. There has been a progressive increase in the total area in the state under mulberry. This phenomenon has largely been due to the adoption of package of programmes by sericulturists for improved mulberry cultivation and rearing of silk-worms developed by the Sericulture Research Organisations. Moreover, a very important fact about irrigation is that the leaf yield of mulberry is nearly 4 to 5 times higher to that of cultivation under rainfed conditions. Adoption of irrigation is bound to increase significantly the leaf yield, the raw silk production and also the income of the farmers.

7.2 Education in Sericulture

In Karnataka, the formal education and research of sericulture came into existence as an offshoot of the department of entomology in the University of Agricultural Sciences, Bangalore. Later in 1974, the University created positions for instructional facilities both at Hebbal and Dharwad campuses in the field of Sericulture. Research activities were also initiated in the Department of Entomology on silkworms as part of Post-graduate research. During 1980, with the assistance of World Bank under Karnataka Sericulture Project-I (KSP-I), M.Sc (Agri.)

in Sericulture was started at the Entomology Department. In 1982, to cater to the trained manpower needs of the State Sericulture Department, a 4-years B.Sc. (Sericulture) degree programme on par with other degrees of the University was started, subsequently one year PG Diploma in Sericulture was also offered simultaneously at the Department. In the year 1986, M.Sc. (Sericulture) and Ph.D. were started in the Department of Sericulture. In the year 1995, the UG Programme was shifted to the newly established Sericulture College at Chintamani. Ever since, the Department has been offering M.Sc. (Seri) and Ph.D. in Sericulture and also offering sericulture courses to B.Sc. (Ag.) and B.Sc. (Ag. Maco) degree programmes of the University. The department is conducting research both by way of M.Sc. and Ph.D. research work and handling many outside funded research projects including basic and applied aspects of Sericulture. Having facilities like established mulberry garden with different commercially cultivated mulberry varieties, mulberry germplasm and host plants of non-mulberry silkworms, grainage and silk reeling units, silkworm pathology laboratory and entrepreneurship development training in cocoon crafts is being imparted to UG as well as PG Students as part of the course curriculum.

7.3 Supply of Graduates

The out turn of graduates and above in sericulture are 34 (2014-15), among them 24 are under-graduates, 7 are post-graduates and 3 are doctoral degree graduates and diploma holders are 15. Looking into past five years data of intake and out turn of graduates in sericulture, the increasing trend exist, in intake 33 in 2010-11 to 37 in 2014-15 with CAGR at 10.01 %, similarly in out turn 18 in 2010-11 to 24 in 2014-15 with CAGR at 10.82 % at under-graduate level. In master's degree, also increasing trend was existed in the intake 5 to 7 (2010-11 to 2014-15) with CAGR 15.74 % and out turn 6 to 7 (2010-11 to 2014-15) with CAGR 0.18 %. Similarly, Doctoral degree intake increased from 2 to 7 (2010-11 to 2014-15) with CAGR 28.47 % and out turn increased 2 to 3 (2010-11 to 2014-15) with CAGR 8.45.00 % (Table 7.1).

Table 7.1: Intake and Out turn of Graduates in Sericulture Courses during 2010-11 to 2014-15

Year	Intake				Outturn				UG and above	
	Diploma	UG	PG	Ph.D.	Diploma	UG	PG	Ph.D.	In	Out
2010-11	10	33	5	2	9	18	6	2	**40**	**26**
2011-12	13	34	5	2	11	21	9	2	**41**	**32**
2012-13	12	38	7	2	12	21	4	2	**47**	**27**
2013-14	15	45	11	2	14	33	11	2	**58**	**46**
2014-15	15	37	7	7	15	24	7	3	**51**	**34**
CAGR (%)	**10.01**	**5.22**	**15.74**	**28.47**	**13.46**	**10.82**	**0.18**	**8.45**	**8.68**	**9.41**

7.4 Areas of Placement for Sericulture Graduates

Number of well qualified sericulture graduates is required in the Sericulture Department, Silk Board, SAU's- teaching, research and extension, Sericulture inputs marketing, chaki rearing, NGO's, Self-Employment and Other sectors.

7.5 Current Stock of Human Resources in Sericulture

There is no direct source of data on the number of active sericulture graduates human resources in the state. Hence, the same procedure was followed as that of section **5.1.4** (Agriculture sector).

An effort was made to estimate the stock of sericulture graduates from annual supply data from college of Sericulture, Chintamani and PG courses offered at UAS, Bangalore and Dharwad. The stock from supply side is estimated from the out turn of graduates in the past 5 years (Annexure-4.1.7). The total stock of graduates and above in sericulture, estimated from annual supplies is about 34 in 2014-15.

7.6 Demand Projections

Sericulture is an agro industry (combining both agriculture and industry), the end-product of which is silk. India has the unique distinction of producing all the four types of silk *viz.*, Mulberry, Muga, Tasar and Eri. About 14 % of the global raw silk is produced in India and Karnataka is the major producer. Besides meeting the ever growing domestic demand, silk in its various forms is also exported and this sector is one of the important foreign exchange earners for the country. It is a labour intensive sector providing direct and indirect employment to a large number of families. Directly and indirectly sericulture is providing jobs for about 10.67 lakh people. One hectare of mulberry provides year long continuous job for 13 persons. Karnataka has well-established Multi-voltine and Bivoltine Seed areas. They cater to the demand of parental seed cocoons required for the production of crossbreed and bivoltine hybrid layings. Almost 88% of Karnataka sericulture is spread in southern part of Karnataka, which is fast modernizing. Factors like urbanization, industrialization, depleting water table, scarcity of agriculture labour have affected sericulture in this part.

Karnataka is one of the five traditional mulberry-silk producing states. Its contribution to the total silk produced in the country is about 48%. This sub-sector provides employment to more than 18.50 lakh people in the state annually.

The graduates are employable in the line department, banking and agricultural financing sectors, private companies, input manufacturing and supply agencies, NGOs, private enterprises dealing with silkworm egg production, Chawki rearing centers (CRCs), silk marketing, import and export of silk and the forward and backward linkages of the enterprises. Besides, the Central Silk Board, Bangalore, being a premier commodity based organization, has its research and extension units spread across the country, have been absorbing the trained sericulture graduates regularly. These graduates have further avenues for continuing higher education within India and abroad.

7.7 Sericulture Department

The Indian raw silk demand is expected to grow at the rate of 5-7% per annum. The domestic consumption of raw silk is also expected to grow at the same rate. The gap between demand and supply is likely to remain in the medium term horizon. In compliance with the National Fiber Policy, Department of Sericulture, Government of Karnataka strives to maintain a balance between the interests of all stakeholders in the value chain, including farmers, reelers, twisters, weavers and value-addition designers. The main objectives are horizontal and vertical expansion of the mulberry area and increase the production and productivity and reduce the production cost. This could be achieved through encouraging contract/corporate farming, market reforms, emphasizing extension activities in new areas in addition to the ongoing programmes. Sericulture is an ideal crop to improve North Karnataka, which has congenial soil, climate, manpower and the opportunity for development of socio-economic conditions.

The data related to the staff positions in the sericulture department have been collected (sanctioned, filled and vacant positions) and estimates have been made on that basis. The estimate showed that there is a 22.56 % vacancy in the state. The number of sericulture officers in the state has been estimated at 1229 (Table 7.2).

Detailed projected stock demand in the sericulture department is shown in Annexure-4.8. The estimated stock requirement in 2014-15 is about 358. This would increase to 583 by 2024-25. Thus, department of sericulture would need about 25 per annum during 2015 to 2025.

Table 7.2: Staff positions in the sericulture department and SAU's as on March 31, 2015

Organisation	Sanctioned	Filled	Vacant	Vacancies %
SAUs	48	27	21	43.75
Department	1587	1229	358	22.56
Total	1635	1256	379	33.15
SAUs- Agriculture (UASB, D, R and S)				
Department: Department of Sericulture, Govt. of Karnataka				

7.8 Summary of Demand Projections

The projections of stocks of sericulture human resources required to the meet the needs of various segments of the economy and the consequential flows required from the educational system are summarized below:

7.9 Stock Projections

The overall stocks of sericulture human resources required to meet the requirements of various segments of sericulture have been added to give projections upto 2024-25. Three sets of projections have been worked out as explained in the segment- level projections – one a high projection, a low projection and an average. These projections are given in Annexure-4.8. The corresponding annual outputs required from the educational system to lead to these stocks are given in **Annexure-4.8.12.**

Considering the current supply and experts opinions on graduates aspirations and employee organizations capacity to absorb graduates, the average variant projection is considered desirable. The overall stock as per the average scenario is given in **Table 7.3.** The stock of graduates and above, would grow from about 31 in 2014-15 to about 464 in

2024-25, *i.e.*, the average addition to the stock during the period 2015 to 2025 is about 35 per annum.

Table 7.3: Overall Projections of Requirements Stocks of Sericulture Human Resources

Year	Diploma*	UG	PG	Ph.D.	All UG & above
2014-15	15	15	12	4	31
2024-25	180	293	140	31	464

* one year course

7.8 Demand – Supply Gap in Stock

In 2014-15, the actual stock from supply is 31. Assuming out turn levels will grow at the same growth rate as observed during the last five years (*i.e.*, 13.22 per cent during 2010-11 to 2014-15), the estimated demand –supply scenario in 2025 will be:

Supply: 104

Demand: 464

Gap: 360 (77.58 % of the demand)

7.9 Flow Projections

The overall annual flow required in all sub-sectors of sericulture is given in Annexure- 4.1.7 and summary in Fig. 7.1.

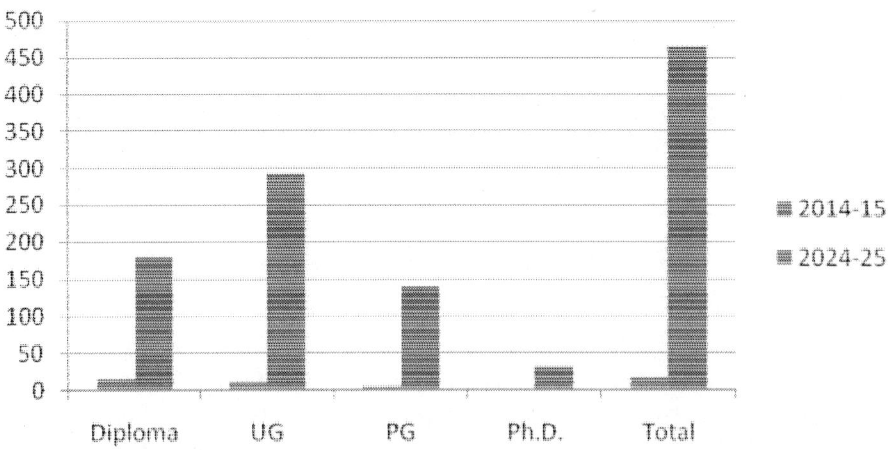

Fig. 7.1: Current (2014-15) and Projected Out turn required (2024-25) in Sericulture

The projections suggest that by 2024-25, the annual out turn from the sericulture education system should be about 180 diploma holders, 293 under-graduates 140 post-graduates and 31 doctorates, rising from 15 diploma holders, 15 under-graduates 12 post-graduates and 4 doctorates in 2014-15.

7.10 Conclusions

Sericulture is one the most important of agriculture and industry sector. In the Southern Karnataka, rural and semi-urban areas, more number of farmers depends on sericulture. Sericulture helps to improve the manufacturing sector. This is an important labour intensive activity and agro-based industry.

Current annual out turn of about 31 graduates is less than the requirement of 464 by 2024-25. There is need to expand sericulture education to produce additionally 474 graduates (31 UG; 14 PG and 1 Ph.D.) per year by 2024-25. This needs expansion and strengthening of the existing college and Department, which offers PG courses. The diploma holders are needed at field level work. The annual supply of diploma holders is about 15 in 2014-15 and the requirement would grow to 60 by 2024-25.

□□□

Chapter 8

Human Resources Assessment in Veterinary & Animal Husbandry Sector

8.1 Introduction

Animal Husbandry plays an important role in the national economy and in the socio-economic development of the country. This sector also plays a significant role in supplementing family income and generating gainful employment in the rural areas, particularly among landless labourers, small and marginal farmers and women, besides providing cheap nutrition to millions of people. The sector ensures better utilisation of human resources, especially in dry and low rain fall areas and this is the sector which has high participation of women. Animal Husbandry is closely inter-connected with agriculture and plays an important complimentary role in the rural economy. The per capita availability of milk and egg during 2010-11 is 226 gms and 50 numbers, respectively as against 216 gms and 47 eggs during 2009-10. There was a steady improvement in milk, meat and egg production during the last decade, which was mainly due to implementation of crossbreeding programme, effective disease diagnosis and control, adoption of new technology, improvement in availability of infrastructural facilities and government programmes and policies.

Livestock provides major additional contribution to agriculture through draft power, manure, and fuel as a fertilizer, animal products such as meat, milk eggs while poultry provide daily cash income and much required nutrition to rural population. Karnataka livestock sector plays an important role in improving the economic status of the rural farmers, who are dependent on the livestock for their livelihood. As per the 19th Livestock Census, 2012, Karnataka has 2.9 crore of livestock and 5.3 crore of poultry population. Government of India has declared the Karnataka state as a render pest disease free zone. To provide health care to the animals and for improvement development of breeds of animals, various programmes were implemented successfully by the Department of Animal Husbandry and Veterinary Services, through its institutional network.

The share of Animal Husbandry in Gross State Domestic Product (GSDP) of Agriculture and Allied activities was 20.51% during 2013-14. The share of Karnataka in all India livestock and poultry population was 5.41% and 7.33%, respectively. The density of livestock in the state was estimated at 51.21 per sq.km and 47,468 per lakh human population. The production of milk in the state was 5.99 million MT during the year 2013-14 and 6.12 million MT for the year 2014-15. During 2014-15, the health of livestock and poultry are rendered through a network of 4,112 Veterinary Institutions, comprising of 364 Veterinary Hospitals, 1943 Veterinary Dispensaries, 1181 Primary Veterinary Centers, 229 Artificial Insemination Centers and 174 Mobile Veterinary Clinics and 221 other Veterinary Institutions.

8.1.2 Education in Veterinary Science and Animal Husbandry

Education in the field of Veterinary Science and Animal Husbandry, though considered allied to agricultural disciplines, is somewhat different. The under-graduate course is of five years duration, whereas the duration is generally four years for all other courses. Unlike graduates from other fields, veterinarians have to register with the Veterinary Council of India (VCI). VCI is a statutory body constituted by the Government of India under the Indian Veterinary Council Act, 1984, to regulate veterinary and animal husbandry practice and education in the country. Only those who possess recognized veterinary qualification and registered can practice in the country. Apart from such registration and maintenance of Indian Veterinary Practitioners Register (IVPR), professional matters like recognition of veterinary teaching institutions, laying down the syllabus for the courses and conducting examinations for admissions against the Central quota of seats in the veterinary colleges are responsibilities that lie within the purview of VCI. In the case of other disciplines, it is the ICAR which discharges some of these functions.

In Karnataka, realizing the importance in growth and development of animal husbandry, dairy and fisheries sector a separate University called Karnataka Veterinary, Animal and Fisheries Sciences University (KVAFSU) was established on 23rd February, 2004. This University is exclusively for the development of education and learning;

conduct of research and extension education and transfer the rural oriented technologies in the areas of Veterinary, Animal, Dairy and Fisheries Sciences in Karnataka.

Animal sciences covered at the KVAFSU include specializations like animal genetics and animal breeding, animal physiology, animal nutrition, fodder technology and related areas. While veterinarians are necessary for treating the diseases of cattle and other animals, human resources in animal science disciplines is needed for research and its applications in animal breeding and animal nutrition to bring about improvement in the quality and yield of the livestock, milk or meat.

At present four colleges are offering courses in veterinary sciences and animal husbandry, two colleges offering Dairy Science courses and one college offering courses in Fisheries in the state. There are two more veterinary sciences and animal husbandry colleges, are being sanctioned by the Govt. of Karnataka and supposed to start the functioning during 2017-18 academic year. The college-wise intake and out turn is given in Annexure-4.1.10 and summary in Table 8.1.

8.1.3 Supply of Graduates

The annual out turn of graduates in veterinary science are 246 (2014-15), among them 31 are diploma holders, 151 are under-graduates, 77 are post-graduates and 18 are doctoral degree graduates. Looking into past five years data of intake and out turn of graduates in veterinary science, the decreasing trend in intake 238 in 2010-11 to 161 in 2014-15 with CAGR at 0.77 %, whereas increasing trend in out turn 102 in 2010-11 to 151 in 2014-15 at under-graduate level with CAGR at 5.65 %. In master's degree, increasing trend existed in the intake and out turn 116 and 49 in 2010-11 to 119 and 77 in 2014-15 with CAGR 0.37 and 12.62 %, respectively. Similarly, increasing trend existed in the intake and out turn of doctoral 17 and 22 in 2010-11 to 12 and 18 in 2014-15 with CAGR 0.33 and 13.46 %, respectively (Table 8.1).

Table 8.1: Intake and Out turn of Graduates in Veterinary Sciences during 2010-11 to 2014-15

Year	Intake				Out turn				UG and above	
	Dip-loma	UG	PG	Ph.D.	Diploma	UG	PG	Ph.D.	In	Out
2010-11	17	238	116	17	15	102	49	12	**371**	**163**
2011-12	45	241	166	85	15	148	79	14	**492**	**241**
2012-13	94	245	133	47	28	162	45	12	**425**	**219**
2013-14	97	236	108	36	29	117	105	22	**380**	**244**
2014-15	97	161	119	22	31	151	77	18	**302**	**246**
CAGR (%)	**52.98**	**0.77**	**0.37**	**0.33**	**23.51**	**5.65**	**12.62**	**13.46**	**0.61**	**8.71**

8.1.4 Areas of Placement for Veterinary Graduates

Graduates with a degree in veterinary enter a broad range of challenging and rewarding professional careers in both public and private sectors. Veterinary Dept., farm universities- R&D and Education, Public Veterinary Services, Private Veterinary Services, Financial institutions, Vet. Pharmaceutical and Animal drugs, Animal Feed Industries, Milk and Meat Processing Industries, Self-Employment, NGOs and other sectors.

8.1.5 Current Stock of Human Resources in Veterinary & Animal Husbandry

There is no direct source of data on the number of active veterinary science human resources in the state. As such, the current stock can only be estimated, as in the case of other agriculture related disciplines, indirectly by cumulating the out turn of graduates, post-graduates and doctorates in veterinary sciences over past five years (as many as the active working life of a graduate, post-graduate or doctor) and adjust for attrition due to retirement, migration and such causes.

An effort was made to estimate the stock of Veterinary & Animal Husbandry graduates from annual supply data from KVAFS, University Bidar. The stock from supply side is estimated from the out turn of graduates in the past 5 years (details given separately in Annexure-4.1.10). The total stock of graduates and above in Veterinary & Animal Husbandry estimated from annual supplies is about 246 in 2014-15.

8.1.6 Demand Projections

The Animal Husbandry sector not only plays an important role in the state economy, and in the socio-economic development of the state/ country, also it contributes significantly in supplementing family incomes and generating gainful employment in the rural sectors, particularly among the landless, small and marginal farmers and women, besides providing cheap nutritional food to millions of people. Though reared mostly on common property resources and crop by products, livestock sector not only ensures food security through providing essential proteins and nutritious human diet in the form of milk, eggs, meat, etc. but also plays an important role in utilization of non-edible agricultural by-products. Livestock provides raw material by-products such as hides and skins, blood, bone, fat, *etc.*

The National Agriculture Policy aims to attain a growth rate in excess of 4% per annum in the agriculture sector, stresses the importance of food and nutritional security issues and the importance of Animal Husbandry sector in generating wealth and employment. Since the present growth rate in crop production is around 2%, higher growth rates of 6 to 8% in Animal Husbandry sector would help in achieving the targetted growth rate of 4% for the Agriculture sector as a whole. The Policy proposes to accord high priority to increasing protein availability in the food basket and generation of exportable surpluses. Health care, fodder production, and freedom from animal diseases are some of the other areas of importance.

The domain of Veterinary services has not remained confined to providing treatment to the sick animals and generating livelihood to the poor, but also takes into account the aspects of animal welfare, food security, food safety, public health and environmental issues, international obligations related to production and health, *etc.* It has, therefore, become imperative to harmonise the standards of education in its entirety as well as strengthen and maintain the standards of practice for ensuring the quality of services in the changed global scenario. As the quality of Veterinary services is essentially dependent on the quality of education, it is necessary that a holistic approach is adopted through a regulatory mechanism for ensuring quality of Veterinary education at all levels. Further, the actions in this direction should match with the increasing expectations from the Veterinarians in terms of enhanced production, productivity and efficient disease control measures to meet the demands of population growth.

8.1.6.1 Normative Approach

The first approach on the basis of number of veterinary institution would depend on the number of animals. This projection is the assumption that the number of animal health care services in the state and growth rate of veterinary institutions. As a first step, enlisting number of veterinary institutions in the state since from 2005-06 (2014-15) and CAGR were calculated (Table 8.3). Projection was made based on CAGR (1.92%) upto 2024-25 **Annexure 4.12.**

The second part of this equivalence concerning is conversion of numbers of animals to livestock unit (Table 8.2). In earlier studies, conducted by IAMR, 2 sheep/goats and pigs were treated as equivalent to one bovine, and 10 poultry were considered equivalent to one bovine.

Table 8.2: Conversion of number of livestock into livestock unit

S.No.	Number of Livestock	Livestock Unit
1	1 Cattle	1
2	1 Buffaloes	1
3	10 Sheep	1
4	10 Goats	1
5	5 Pigs	1
6	100 Poultry birds	1
7	5 Others	1

As regards the number of animals per veterinarian, the norms of recommendation of the National Commission on Agriculture (1976), there should be at least one veterinarian for every 5000 cattle units. The corresponding projections for the stock of veterinarians required by 2024-25, would be 4964 against 4112 in 2014-15. These projections have also been shown in **Annexure- 4.12.1**. Based on livestock population, presently 3006 veterinary institutions are required to meet the demand of the sector (**Table 8.3**). At present, the numbers of veterinary institutions are more than required in the state. However, veterinary institutions are not in accordance with livestock population.

Table 8.3: Trends in Number of Veterinary Institutions (VI) in Karnataka

S.No.	Year	No. of VI
1	2005-06	3238
2	2006-07	3374
3	2007-08	3411
4	2008-09	3680
5	2009-10	3677
6	2010-11	3677
7	2011-12	3680
8	2012-13	3680
9	2013-14	3680
10	2014-15	4110
	CAGR %	**1.92**

8.1.6.2 Based on Sectoral Growth Approach

The second approach adopted for human resources projections is that followed for other agricultural disciplines and considers the requirements according to the current levels and growth patterns in different segments of the veterinary sector as in Parnes model. The segments considered are: (1) Public Veterinary Services; a) Developmental Departments, b) University-Teaching, Research and Education, c) KMF (2) Private; a) Animal Health Services, b) Animal Feed Industries breeding and fodder, c) Veterinary Pharmaceuticals, d) Milk and Meat processing industries, e) Poultry Industries and f) Other sectors.

8.1.6.2.1 Public Veterinary Services

The department of Animal Husbandry & Veterinary Services started functioning independently from 1945. The department is playing a vital role in providing healthcare services to livestock, extension and training to farmers, implementation of socio-economic programs, conducting Sample Survey and Animal Census programs periodically in the State.

Veterinary Institutions across the State are divided into four categories in order to provide better system of services: Veterinary Hospital, Veterinary Dispensary, Mobile Veterinary Service, Primary

Veterinary Centre, Departmental Veterinary Institutions are well-equipped and having trained staff. Along with treating the ailing animals, vaccination against various diseases is carried out as a preventive measure. Mobile Veterinary Service operating in the State is providing veterinary healthcare services to the livestock of those farmers living in villages, where veterinary institutions are not present.

In order to achieve the target of the department of Animal Husbandry & Veterinary, there is lot of demand for skilled veterinary graduates in the state. The data related to the staff positions in the Animal Husbandry & Veterinary department have been collected (sanctioned filled and vacant positions) and estimates have been made on that basis. The estimate showed that there is a 26.28 % vacancy in the state. The number of veterinary officers/doctors in the state has been estimated at 2400 (Table 8.4).

Detailed projected stock demand in department of Animal Husbandry & Veterinary is shown in Annexure-4.12.1. The estimated stock requirement in 2014-15 is about 878. This would increase to 1070 by 2024-25. Thus, department would need about 30 veterinarians per annum during 2015 to 2025.

8.1.6.2.2 Veterinary/Agriculture Universities

Different agricultural/veterinary universities in the state employ veterinary post-graduates/doctoral for distinct posts from concerned field of their specialization. However, for a number of posts of associate professor, assistant professor and other teaching, research and extension are vacant. To meet the technical manpower required for the veterinary sector in the State, The Karnataka Veterinary, Animal and Fisheries Sciences University, Bidar, came into existence to provide teaching, research and extension education services related to veterinary and animal sciences, dairy sciences, fisheries and other allied sciences. Undertaking the extension education of such science and technologies, specially for the rural people of the State. The Veterinary University including other farm universities (Teaching, research and extension) had a combined sanctioned strength of 395 faculties, of which 210 positions (46.84 %) were vacant as on March 31, 2015 details were given in (Table 8.4).

Detailed projected stock demand in Universities is shown in Annexure-4.12.2 The estimated stock requirement in 2014-15 is about 185. This would increase to 301 by 2024-25. Thus, SAU's including

Teaching, Research and Education; need about 15 per annum during 2015 to 2025.

Table 8.4: Staff positions in the Animal Husbandry & Veterinary department and Universities as on March 31, 2015

Organisation	Sanctioned	Filled	Vacant	Vacancies %
Universities	395	210	185	46.84
Department	3278	2400	878	26.78
Total	**3673**	**2610**	**1063**	**36.81**
Universities – Veterinary and Other				
Department: Animal Husbandry & Veterinary Department, Govt. of Karnataka				

8.1.6.2.3 Karnataka Milk Federation (KMF)

Dairy industry, particularly the co-operative sector, also engages the services of veterinarians and animal nutritionists to enhance the quality and productivity of the milch cattle. KMF recruit about 12 to 15 veterinary graduates in year since it has substantial programmes of animal husbandry and animal health. The total dairy sector utilization of veterinary human resources may be placed at about 221 graduates now, which may go upto 360 graduates (a growth of 5% per annum) by 2024-25 in view of the importance of animal health and quality improvement for the ambitious dairy development programme.

Detailed projections are given at Annexure 4.12.4. The stock requirement of veterinary human capital for dairy industry would grow from 221 in 2014-15 to about 360 by 2024-25, *i.e.* about 10 per annum during the period 2014-15 to 2024-25.

8.1.6.2.4 Poultry Sector

The poultry industries with a significant presence of veterinarians are becoming very much essential to achieve targetted growth. Poultry farming in Karnataka has transformed into a techno-commercial industry from the status of backyard farming since three decades. Poultry consists of generally three categories, namely Fowls, Ducks and Turkey & others. The Total poultry population in the state is 5,34,42,030 numbers. Karnataka stands as fourth largest egg producer and fifth chicken meat producer in the Country with about 27,352 lakh eggs and 6337 tonnes of chicken meat in 2014-15. This production is achieved generally by commercial poultry operations; however, a significant contribution comes from rural poultry also. Keeping this in background, the Government of Karnataka formulated policies to support and augment rural poultry. This was planned and being implemented through central sector schemes with 3- tier structure. The first tier involves Central government institutions, ICAR organizations, State Agriculture Universities, Veterinary Universities and private rural bird breeders for supplying the requisite parent stock of rural birds. The second tier involves State Institutions, State Poultry Farms, and District Rearing Centers, *etc.*, which were duly assisted by a scheme called "Assistant to State Poultry Farms" for strengthening infrastructure and to meet operational expenditure. The third tier involves grass root institutions, NGOs and farmers to operationalize the scheme.

Data on number of the veterinary graduates employed by the poultry industry were collected by discussion with experts and through establishment survey. There are about 20 veterinary graduates engaged in poultry sector in 2014-15 and this may increase upto 33 with 5 % growth rate by 2024-25, **Annexure 4.12.3.**

8.1.6.2.5 Veterinary Pharmaceuticals

With the rapid growth in dairy and poultry industry, animal health has assumed considerable importance. There are about 5 pharmaceutical firms, engaged in production and distribution of animal drugs and other medical equipment for treating animals. The pharmaceutical companies engage not only veterinarians and animal scientists but bio-technologists and pharmacy graduates also for their research and development jobs. Establishment survey showed that, on an average, a veterinary pharmaceutical company engaged 4 veterinarians (2 graduates, 1 post-graduates and 1 doctorate), the total number would be 20. The pharmaceutical industry has been presently growing at about 10 per cent per annum. Therefore, the number of veterinarians and animal scientists is likely to increase at least at a rate of 8 per cent per annum to about 43 by 2024-25. Breakup by qualifications has been derived with data from establishment survey.

Detailed projections are at Annexure- 4.12.6.**The requirement of human capital for veterinary pharmaceutical industry would grow from 20 in 2014-15 to about 43 by** 2024-25. The requirement during 2010-20 is about 2 per annum.

8.1.6.2.6 Animal Feed Industry

Animal feed is an important input to enhance animal quality. According to some estimates, there is a supply demand gap in fodder production – to the extent of over 60 per cent in the case of green fodder and about 25 per cent in the case of dry fodder. Only 50 per cent of the marketed feed is produced in the organized sector. There are thousands of animal feed producers in the country/state; in Karnataka, about 14 major animal feed industries are there. As per report of IAMR indicated that on an average, a feed plant employed 30 persons, of whom about 7 comprise of human resources relating to agricultural sciences. About 2 of them are graduates and post-graduates in animal sciences and veterinary fields and the rest are diploma holders. On this basis, and assuming that the operating factories now number about 14, it is estimated that the total human resources in animal and veterinary sciences in the animal feed units at present (2014-15) is of the order of 28 (graduates and above). Projections for the future have been obtained assuming that animal feed production will grow at a rate of 8 per cent per annum, which was the rate of growth of total employment in animal feed industry during the period 2004-05 to 2014-15 (Annual Survey of Industries, 2015).

Detailed projections for animal feed sector are given at Annexure 4.12.7. The stock requirement of human capital would grow from 28 in 2014-15 to about 60 by 2024-25 with an average of 4 per annum.

8.1.6.2.7 Meat Processing

Meat sector plays an important role, as it not only provides meat and by-products for human consumption but also contributes towards sustainable livestock development and livelihood security for millions of men and women from weaker sections. Major portion of meat from sheep, goat, pig and poultry is primarily used for domestic consumption in the form of hot meat. Certain portion of meat from buffaloes, cattle and sheep is exported. There are, however, 8 to 10 major meat processing units in Karnataka and employment in this sub-sector had grown at an annual rate of 10 per cent in the last decade. As per IAMR report, about 3 per cent of this employment comprises agri-human resources. It is around 1 per processing plant.

Projections upto 2025 have been made on this basis, and are shown in Annexure 4.12.5. The projected stock demand for 2024-25 is 26, which would increase from 10 in 2014-15. The annual requirement of veterinarians in meat processing sector is about 1 per annum.

8.1.6.2.8 Private Veterinary Services

Private veterinary services include those provided by non-governmental organizations and also individual private veterinary practitioners. Private veterinary practice is still in its infancy in Karnataka and by and large confined to urban and city areas, and is undertaken by veterinarians, who retired from government service. There are no precise estimates of the number of veterinarians providing private services. According to various estimates available, the number of private practitioners have been growing over the years. As per NAARM report, they estimated that number of private veterinary practitioners would be about 3,000 in 2008, 5,000 in 2015 and 8,000 in 2020, in India. Apart from private veterinary practitioners, there are NGOs operating livestock development programmes. BAIF Development Research Foundation has a Central Research Station, cattle breeding and Semen Freezing Laboratory and nation-wide network of centres providing livestock improvement services such as artificial insemination. There are also Goshalas operated in the non-government sector, which eventually will require veterinary services.

Projections for private veterinary services have been shown in Annexure 4.12.8. **The requirement of human capital would grow from 100 in 2014-15 to about 163 by 2024-25. The requirement during 2015-25 would be about 8 per annum.**

8.1.6.2.9 Others

The human resources requirements of the sectors covered above have been inflated by 10 per cent to take care of any other sectors not specifically covered.

8.1.7 Summary of Demand Projections

The projections of stocks of veterinary and animal husbandry human resources required to meet the needs of various segments of the economy and the consequential flows required from the educational system are summarized below:

8.1.8 Stock Projections

The overall stocks of veterinary and animal husbandry human resources required to meet the requirements of various segments of dairy sector have been added to give projections upto 2024-25. Three sets of projections have been worked out, as explained in the segment-level projections - one a high projection, a low projection and an average. These projections are given in Annexure-4.12. The corresponding annual outputs required from the educational system to lead to these stocks are given in **Annexure-4.12.12.**

Considering the current supply and experts opinions on graduates aspirations and employee organizations capacity to absorb graduates, the average variant projection is considered desirable. The overall stock as per the average scenario is given in **Table 8.5.** The stock of graduates and above would grow from about 415 in 2014-15 to about 681 in 2024-25, *i.e.,* the average addition to the stock during the period 2015 to 2025 is about 35 per annum.

Table 8.5: Overall Projections of Veterinary and Animal Husbandry Human Resources

Year	UG	PG	PhD	All UG & above
2014-15	151	77	18	246
2024-25	971	502	224	1697

8.1.9 Demand – Supply Gap in Stock

In 2014-15, the actual stock from supply is 246. Assuming out turn levels will grow at the same growth rate as observed during the last five years (*i.e.,* 8.71 per cent during 2010-11 to 2014-15), the estimated demand-supply scenario in 2025 will be:

Supply: 551

Demand: 1697

Gap: 1146 (67.53 % of the demand)

8.1.10 Flow Projections

The overall annual flow required in all sub-sectors of veterinary and animal husbandry is given in Annexure-4.1.10 and summary in Fig. 8.1.

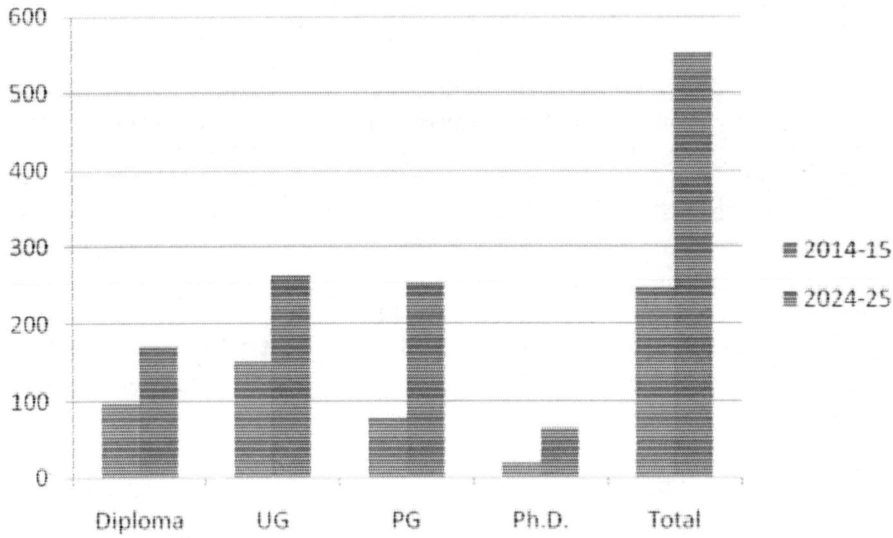

Fig. 8.1: Current (2014-15) and Projected Out turn required (2024-25) in Veterinary and Animal Husbandry

The projections suggest that by 2024-25, the annual out turn from the veterinary and animal husbandry education system should be about 971 under-graduates, 502 post-graduates and 224 doctorates, rising from 151 under-graduates, 77post-graduates and 18 doctorates in 2014-15.

8.2 Human Resource Assessment in Dairy Sector

Karnataka state has always remained in the forefront of all agricultural development initiatives in the country and dairy development is not an exception. At a time, when agriculture was grappling with the consequences of rapid and unidirectional intensification in its production process, dairy farming provided not just the much-required alternative but also revived and further strengthened the animal-crop production linkage. Taking full advantage of the potential of dairy enterprise to provide continuous and regular employment as well as income, production in the state increased rapidly. Dairy farming in Karnataka, as elsewhere in the country, is largely characterized by the prevalence of dairy enterprises that are mostly subsidiary occupations alongside the main agriculture activity of the farmers. Specialized dairy enterprises do exit but not only their number abysmally low as compared to regular types but also are restricted mostly to urban areas and their surroundings.

Dairy development in Karnataka has had a long and conquered history, the growth in population of dairy animals, as well as the production substantiates it. It all started with the setting up of imperial Institute of Animal Husbandry and Dairying in 1923 to serve as the Centre of Dairy training and Research. Later this centre became the Indian Dairy Research Institute; and then the Southern Region Station of the National Dairy Research Institute at Bangalore.

Bangalore Dairy was commissioned in 1965 with UNICEF aid with capacity of 50,000 litres of milk per day. By the year 1973, Karnataka State had 10 Dairy Plants with a processing capacity of 1.67 lakh litres per day. To begin with, most of these dairies were under-utilized and the milk handled was not even 50 per cent of their capacity. Karnataka Dairy Development Corporation was set up to implement the dairy development project funded by World Bank in the year 1974. The Karnataka Milk Products Limited, which was established in the year 1980, was transferred to KMF in the year 1984. In the year 1985, Government transferred the dairies to the control of KMF. The successive phase of Operation Flood Programme started during 1987, continued until April 1996, covering the entire state through 13 milk unions. At present-KMF is the third largest co-operative organisation in the co-operative sector in the country in terms of procurement and sale. It stands first in the South India. Thus, dairy farming has been developing very fast in Karnataka since 1974.**8.2.1 Education in Dairy Technology.**

Dairy Technology refers to the study of milk and milk-derived food products from a food science perspective. It focuses on the biological, chemical, physical, and microbiological aspects of milk itself, and on the technological (processing) aspects of the transformation of milk into its various consumer products, including beverages, fermented products, concentrated and dried products, butter and ice cream. Milk is as ancient as mankind itself, as it is the substance created to feed the mammalian infant. All species of mammals, from man to whales, produce milk for this purpose. Fermented products such as cheeses were discovered by accident, but their history has also been documented for many centuries, as has the production of concentrated milks, butter, and even ice cream. Technological advances have only come about very recently in the history of milk consumption, and our generations will be the ones credited for having turned milk processing from an art to a science. The availability and distribution of milk and milk products, today in the modern world is a blend of the centuries old knowledge of traditional milk products with the application of modern science and technology.

The separate college of Dairy Technology was started in Karnataka by looking to the need for Dairy Professionals, during 1980 at Bangalore. Later one more college started at Gulbarga during 2007. The dairy technology education has focused research on (i) value addition to dairy products through use of membrane processed dairy ingredients, and (ii) developing Mozzarella cheese analogue based on casein(s) and vegetable/milk fat.

8.2.2 Supply of Graduates

The annual out turn of graduates in Dairy Technology are 47 (2014-15), among them 38 are under-graduates, 7 are post-graduates and 2 are doctoral degree graduates. Looking into past five years data of intake and out turn of graduates in Dairy Technology, the increasing trend exist, in intake 54 in 2010-11 to 59 in 2014-15 with CAGR at 2.15 %, similarly in out turn 31 in 2010-11 to 38 in 2014-15 with CAGR at 9.61 % at under-graduate level. In master's degree, also increasing trend was existed in the intake 13 to 20 (2010-11 to 2014-15) with CAGR 1.00 % and out turn 3 to 7 (2010-11 to 2014-15) with CAGR 11.42 %. Similarly, in Doctoral degree intake 1 to 2 (2010-11 to 2014-15) with CAGR 14.87 % and out turn 1 to 2 (2010-11 to 2014-15) with CAGR 7.18 % (Table 8.6 and Annexure-4.1.12).

Table 8.6: Intake and Out turn of Graduates in Dairy Technology Courses during 2010-11 to 2014-15

Year	Intake			Out turn			UG and above	
	UG	PG	Ph.D.	UG	PG	Ph.D.	Adm	Pass
2010-11	54	13	1	31	3	1	**68**	**35**
2011-12	55	30	3	24	24	2	**88**	**50**
2012-13	62	26	3	31	17	1	**91**	**49**
2013-14	57	14	3	40	13	1	**74**	**54**
2014-15	59	20	2	38	7	2	**81**	**47**
CAGR (%)	**2.15**	**1.00**	**14.87**	**9.61**	**11.42**	**7.18**	**1.78**	**6.89**

8.2.3 Areas of Placement for Dairy Technology Graduates

Graduates with a degree in Dairy Technology can enter a broad range of challenging and rewarding professional careers in public, private and co-operative sectors. In co-operative sector, KMF & Co-operative societies, in Public sector, Universities and state development departments and in private, milk and milk processing industries, Drug Industries and Lab, Animal feed industries, NGOs and others (Annexure-2).

8.2.4 Current Stock of Human Resources in Dairy Technology

There is no direct source of data on the number of active Dairy Technology human resources in the state. As such, the current stock can only be estimated, as in the case of other agriculture related disciplines, indirectly by cumulating the out turn of graduates, post-graduates and doctorates in Dairy Technology over past five years (as many as the active working life of a graduate, post-graduate or doctor) and adjust for attrition due to retirement, migration and such causes.

An effort was made to estimate the stock of Dairy Technology graduates from annual supply data from different colleges. The stock from supply side is estimated from the out turn of graduates in the past 5 years. The total stock of graduates and above in Dairy Technology, estimated from annual supplies is about 47 in 2015.

8.2.5 Demand Projections

Dairy Technology is growing field of agriculture that deals with the milk production processing and value addition for that lot of skilled human resources are needed.

8.2.5.1 Karnataka Co-operative Milk Producers' Federation Limited (KMF)

Karnataka Co-operative Milk Producers' Federation Limited (KMF) is the Apex Body for the dairy co-operative movement in Karnataka. It is the second largest dairy co-operative amongst the dairy cooperatives in the country. In South India, it stands first in terms of procurement as well as sales. One of the core functions of the Federation is marketing of Milk and Milk Products. The Brand "Nandini" is the household name for Pure and Fresh milk and milk products. KMF has 14 Milk Unions covering all the districts of the State, which procure milk from Primary Milk Co-operative Societies (PMCS) and distribute milk to the consumers in various Towns/Cities/Rural markets in Karnataka.

KM F has responsibility providing remunerative price and market to the rural milk producers of the state and supplying pure milk and milk products to the consumers. The KMF has 22 dairy processing plants with a capacity of 56.80 lakh liters /day. This organization has 44 chilling centers to process 21.40 lakh liters of milk and also 5 milk product dairies, which manufacture 100 MT milk powder every day. The KMF in its jurisdiction has 5 cattle feed plants, which produces 45,000 MT's of cattle feed/month, the plant has ISO 9001/2000 certification for quality production and supply of cattle feed to producers. At present, 13,207dairy co-operative societies are functioning, 23.38 lakh farmers are enrolled as members, out of which, 7.98 lakh farmers are active members.

In order to achieve the target of the KMF, there is a greater demand for skilled dairy technology graduates in the state. The number of dairy technology graduates required in the state was estimated by taking the expert's opinion and establishment survey (Table 8.7).

Detailed projected stock demand in KMF is shown in Annexure-4.10.1.The estimated stock requirement in 2014-15 is about 104. This would increase to 169 by 2024-25. Thus, KMF would need about 10 per annum during 2015 to 2025.

8.2.5.2 Universities/Colleges

The veterinary/dairy/agricultural colleges in the state employ dairy technology post-graduates/doctoral for distinct posts from concerned field of their specialization. The Universities/colleges (including all teaching, research and extension) had a combined sanctioned strength of 63 faculties, of which 41 positions (34.92 %) were vacant as on March 31, 2015 details was given in (Table 8.7).

Detailed projected stock demand in farm universities is shown in Annexure-4.10.2 The estimated stock requirement in 2014-15 is about 22. This would increase to 36 by 2024-25. Thus, SAU's including Teaching, Research and Education; need about 2 per annum during 2015 to 2025.

Table 8.7: Staff positions in the KMF and dairy technology colleges as on March 31, 2015

Organisation	Sanctioned	Filled	Vacant	Vacancies %
SAUs	63	41	22	34.92
KMF	-	104	-	-
University- Veterinary/dairy colleges				
KMF- Karnataka Cooperative Milk Producers' Federation Limited				

8.2.5.3 Private Sector Employment

Dairy technology graduates can become a dairy technologist in various private companies *viz.*, milk and milk processing industries, Drug Industries and Lab, Animal feed industries, NGOs and others. Dairy technology graduates have contemporary, challenging career in private sectors. As per establishment survey and discussion with experts, there are about 25 graduates in milk and milk processing industries, 6 graduates in Drug Industries and Lab, 20 graduates in Animal feed industries, 13-15 graduates in NGOs and others.

On the basis of the data in Annexures-4.1.3, projections have been made of the Dairy technology resources needed till 2024-25 in private sector. **It may be observed from these projections that the private sector is expected to employ 120 by 2024-25. The stock requirement of human capital would grow about 6-8 per annum during the period 2024-25** (milk and milk processing industries, Drug Industries and Lab, Animal feed industries) Tables-8.8, 8.9 and 8.10.

8.2.5.4 NGOs

The technological advancement in the dairy science makes many changes in the milk production. In order to transfer the various technologies to the farmers to improve profitability, milk quality, lifestyle and animal welfare, various NGO's came into existence. Presently, five major NGO's involved in the dairy Technology related aspects. Approximately, one dairy technology graduates required per NGO,s means in 2014-15 about 5 dairy technologist are required this may increased upto 6-7 in 2024-25 (Table 8.11).

8.2.5.5 Others

The human resources requirements of the sectors covered above have been inflated by 10 per cent to take care of any other sectors not specifically covered (Table 8.12).

Table 8.8: Projected Dairy Technology Human resource requirement in milk and milk processing industries till 2024-25

Year	Diploma	UG	PG	Ph.D.	Total
2014-15	18	7	0	25	18
2024-25	19	8	0	27	19

Table 8.9: Projected Dairy Technology Human resource requirement in Drug Industries and Lab till 2024-25

Year	Diploma	UG	PG	Ph.D.	Total
2014-15	4	2	0	6	4
2024-25	4	2	0	6	4

Table 8.10: Projected Dairy Technology Human resource requirement in Animal feed industries till 2024-25

Year	Diploma	UG	PG	Ph.D.	Total
2014-15	9	11	1	20	9
2024-25	9	11	1	20	9

Table 8.11: Projected Dairy Technology Human resource requirement in NGO's till 2024-25

Year	Diploma	UG	PG	Ph.D.	Total
2014-15	4	1	0	5	4
2024-25	4	1	0	5	4

Table 8.12: Projected Dairy Technology Human resource requirement in other sectors till 2024-25

Year	Diploma	UG	PG	Ph.D.	Total
2014-15	6	2	0	8	6
2024-25	6	2	0	8	6

8.2.6 Summary of Demand Projections

The projections of stocks of dairy technology human resources required to meet the needs of various segments of the economy and the consequential flows required from the educational system are summarized below:

8.2.7 Stock Projections

The overall stocks of dairy human resources required to meet the requirements of various segments of dairy sector have been added to give projections upto 2024-25. Three sets of projections have been worked out, as explained in the segment- level projections – one a high projection, a low projection and an average. These projections are given in Annexure-4.10. The corresponding annual outputs required from the educational system to lead to these stocks are given in **Annexure-4.10.11.**

Considering the current supply and experts opinions on graduates aspirations and employee organizations capacity to absorb graduates, the average variant projection is considered desirable. The overall stock as per the average scenario is given in **Table-8.13.** The stock of graduates and above, would grow from about 415 in 2014-15 to about 681 in 2024-25, *i.e.*, the average addition to the stock during the period 2015 to 2025 is about 35 per annum.

Table 8.13: Overall Projections of Dairy Technology Human Resources

Year	UG	PG	Ph.D.	All UG & above
2014-15	38	7	2	47
2024-25	189	98	29	316

8.2.8 Demand – Supply Gap in Stock

In 2014-15, the actual stock from supply is 47. Assuming out turn levels will grow at the same growth rate as observed during the last five years (*i.e.*, 6.89 per cent during 2010-11 to 2014-15), the estimated demand-supply scenario in 2025 will be:

Supply: 120

Demand: 306

Gap: 186 (60.31 % of the demand)

8.2.9 Flow Projections

The overall annual flow required in all sub-sectors of Dairy Technology is given in Annexure-4.1.12 and summary in Fig. 8.2.

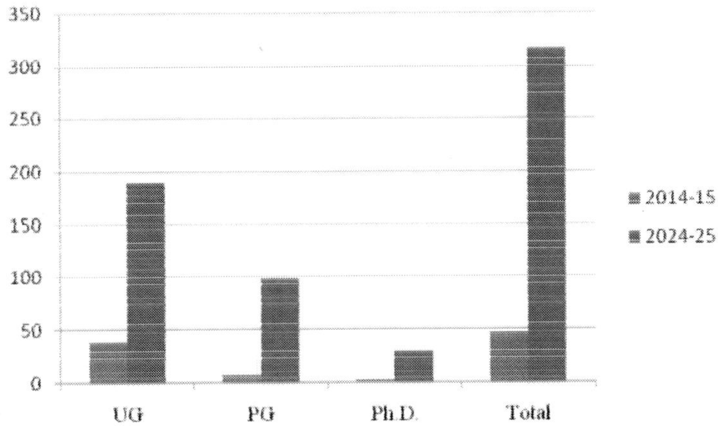

Fig. 8.2: Current (2014-15) and Projected Out turn Required (2024-25) in Dairy Technology

The projections suggest that by 2024-25, the annual out turn from the Dairy Technology education system should be about 189 under-graduates, 98 post-graduates and 29 doctorates, rising from 38 under-graduates, 7 post-graduates and 2 doctorates in 2014-15.

8.3 Conclusions

Looking into the demand for veterinary graduates, the education system need to stabilize and consolidate the higher level education and the intake at UG level must increase. In addition to existing colleges, the efforts are made to establish two more colleges in the state to meet the requirements. It is also noted that the existing colleges are not having sufficient faculty to teach as per the guidelines. Thus, further expansion of education in this sector needs firm commitment on investment towards infrastructure so as to meet requirements. There is a great demand of functional skills among the students that are coming out of educational system.

Current annual out turn of about 246 graduates is less than the requirement of 1697 by 2024-25. There is need to expand veterinary education to produce additionally 1697 graduates (97 UG; 50 PG and 22 Ph.D.) per year by 2024-25. The diploma holders are needed in rural areas for veterinary service because diploma holders would be preferred in place of general graduates now being employed by many employers. The annual supply of diploma holders is about 97 in 2014-15 and the requirement would grow to 168 by 2024-25.

❑❑❑

Chapter 9

Human Resource Assessment in Fisheries Sector

9.1 Introduction

The importance of the fisheries sector to national economy is widely recognised. Its significance lies in three main areas: (1) as a source of animal protein for human consumption, (2) as a source of employment, and (3) as a source of foreign exchange earnings. Fisheries sector occupies a very important place in the overall socio-economic development of the Karnataka. Most importantly, it is the source of livelihood for a large section of economically backward population of coastal districts as well as others districts in the state. It has been recognized as a foremost income and employment generator, as it stimulates growth of a number of subsidiary industries and is a source of cheap and nutritious food besides being a foreign exchange earner. In the year 2015, the total fisher folk population of the state was 8.72 Lakhs comprising 3.11 Lakhs people depending on marine for their livelihood and 5.61 Lakhs people depending on inland fisheries.

9.2 Education in Fishery Science

The College of Fisheries, Mangalore a premier Fisheries Education and Research Institute was established in 1969 with under-graduate degree programme in Fishery Science by the then University of Agricultural Sciences, Bangalore. The College has now become part of Karnataka Veterinary, Animal and Fisheries Sciences University, Bidar. The mission of the College is to promote a greater understanding and appreciation of the biological, technical and economic importance of fisheries and related areas.

The College has Microbiology, Fish Processing Technology, Fisheries Resources and Management and Fisheries Environment and Ecology, Fishery Processing and Engineering Divisions.

Table 9.1: The post-graduate and doctoral degree programmes in the faculty of Fisheries Sciences

S.No.	Programmes	Year of start
	I. M.F.Sc. (Duration 4 Semester)	
1	Aquaculture	1974
2	Fisheries Resources & Management	1974
3	Aquatic Environment Management	1974
4	Fisheries Microbiology	1987
5	Fish Processing Technology	1987
6	Fisheries Engineering & Technology	1988
	II. Ph.D. (Fisheries Sci.) (6 Semester)	
1	Aquaculture	1980
2	Fish Processing Technology	1980
3	Fisheries Resources & Management	1980
4	Aquatic EnvironmentManagement	1991
5	Fisheries Microbiology	1991

There are total six post-graduate (Aquaculture, Fisheries Resources & Management, Aquatic Environment Management, Fisheries Microbiology, Fish Processing Technology and Fisheries Engineering & Technology) and five doctoral (Aquaculture, Fish Processing Technology, Fisheries Resources & Management, Aquatic Environment and Fisheries Microbiology) degree programmes are being offered at college of fisheries (Table 9.1).

9.3 Supply of Graduates

In Karnataka, fishery education and research have been accorded fairly high priority and in order to provide elaborated institutional set-up fishery college was established in Mangalore in the year 1969 with under-graduate degree programme with intake capacity of 22. To meet the gap qualified human resources in education and research of fisheries science, the post-graduation and doctoral degree programmes were started in year 1974 and 1980, respectively.

The annual out turn of graduates in fishery sciences are 39 (2014-15), among them 22 are under-graduate, 15 are post-graduate and 2 are doctoral degree graduates. Looking into past five years data of intake and out turn of graduates in fisheries science, the decreasing trend exist in intake 45 in 2010-11 to 32 in 2014-15 with CAGR at 0.91%, whereas increasing trend in out turn 16 in 2010-11 to 22 in 2014-15 at under-graduate level with CAGR at 18.36%. In master's degree, increasing trend existed in the intake and out turn 8 and 19 with CAGR 18.36 % in 2010-11 to 16 and 22 with CAGR 18.67% in 2014-15, respectively. There was decreasing trend in the intake of doctoral graduates 7 to 3 (2010-11 to 2014-15) with CAGR 0.25, whereas increasing trend in out turn of doctoral 2 to 6 (2010-11 to 2014-15) with CAGR 24.57% (Table 9.2).

Table 9.2: Intake and Out turn of Graduates in Fishery Science Courses during 2010-11 to 2014-15

Year	Intake			Out turn			UG and above	
	UG	PG	Ph.D.	UG	PG	Ph.D.	In	Out
2010-11	45	8	7	16	10	2	**60**	**32**
2011-12	45	23	14	14	13	1	**82**	**27**
2012-13	44	11	10	20	7	4	**65**	**31**
2013-14	34	22	8	41	9	1	**64**	**51**
2014-15	32	19	3	22	15	6	**54**	**39**
CAGR (%)	**0.91**	**18.36**	**0.25**	**18.67**	**4.53**	**24.57**	**0.44**	**10.86**

9.4. Areas of Placement for Fisheries Graduates

A number of sectors of the economy have openings for personnel qualified in fisheries sciences. Some of the important sectors are:

9.4.1 Public Sector

State Fisheries Department, Fisheries College and Agriculture Research Institutes, different organization under the central government *viz.*, Indian Council of Agricultural Research, Marine Products Export Development Agency, Fishery Survey of India(FSI), Fish Farmers Development Agencies (FFDAs), Coastal Aquaculture Authority (CAA), National Centre for Sustainable Coastal Zone Management (NCSCZM), National Fisheries Development Board, Marine Product Export Development Authority (MPEDA), Indian National Centre for Ocean Information Service, Kisan Call Centre, Financial Institutions, *etc.*

9.4.2 Private Sector

Processing plants, Fish/Shrimp seed hatcheries, Shrimp/Fish rearing units, Feed mills, Canning industries, aquaculture farms, aqua-feed plants, deep sea fishing vessels, and pharmaceutical companies, NGOs engaged in activities relating to fisheries development, *etc*..

9.4.3 Entrepreneurship

For people interested in entrepreneurship, there is sufficient technical guidance, financial assistance and subsidies provided by the National Fisheries Development Board, Agri-clinic and Agri-Business Centre, NABARD and other nationalized banks. NABARD gives financial support to the candidate having B.F.Sc. degree, for fishery-related self-employment enterprises. The main areas for entrepreneurs in Karnataka are ornamental fish culture, aquaculture, hatchery and seed production, commercial pearl production, fish disease diagnostic centre, consultancy services and establishment of agri-clinics.

9.4.4 NGO's

For people who are oriented more towards community mobilization and social upliftment of Fishers, they can join Non-Governmental Organization.

9.4.5 Other Ancillary Sectors

There are a number of activities that support fishing industry. Net mending, boat building, and manufacturing of fishing accessories, aqua feed preparation, gelatin, fish oil antibiotic and aqua drug supply are some such industries associated with the fisheries sector. Fishing by trawlers in high seas, preservation and marketing of sea products also need a variety of trained personnel.

9.5 Current Stock of Fishery Science Personnel

There is no direct source of data on the number of active fisheries science graduates in the state. As such, the current stock can only be estimated, as in the case of other agriculture related disciplines, indirectly by cumulating the out turn of graduates, post-graduates and doctorates in Fishery sciences over past five years (as many as the active working life of a graduate, post-graduate or doctor) and adjust for attrition due to retirement, migration and such causes.

Central Institute of Fisheries Education (CIFE), Mumbai in collaboration with Fisheries College Mangaluru, in its previous study on Fisheries education in Karnataka: Trends, current status, performance and prospectus, had estimated the stock of graduates in fisheries sciences as on 1972 to 2014 (Vinay *et al.*, 2016). A similar situation was observed in the secondary data collected for the present study (Table 9.2).

An effort was made to estimate the stock of fisheries science graduates from annual supply data from Fisheries College Mangaluru. The stock from supply side is estimated from the out turn of graduates in the past 5 years. The total stock of graduates and above in fisheries science, estimated from annual supplies is about 39 in 2015.

9.6 Demand Projections

Fisheries in the national economy have gained importance both as a foreign exchange earner and as a generator of employment. In addition, this sector has assumed the characteristics of an industrial enterprise. Fish is the cheapest source of protein available for human diet.

9.6.1 Public sector

In State Governments, job opportunities exist in fisheries department for fisheries graduate for the post of Assistant Fisheries Development Officer (AFDO)/Fisheries Extension Officer (FEO) and District Fisheries Development Officer, financial institutions, and research, teaching & extension at many colleges and universities.

9.6.1.1 Development Departments

The Department of Fisheries was established in the year 1957 with an object to give fillip for production of fish by utilising the rich resources in Marine and Inland sector. Karnataka has about 5.93 lakh ha of inland water resources, 8000 ha of brackish water and 320 km of coastline with 27,000 Sq.km of continental shelf. The department is implementing several schemes for the better utilization of the resources and for the welfare of fishermen in particular and public in general. The fish production in Karnataka was around 2.0 lakh tonnes in early eighties and reached to a peak of over 3.0 lakh tonnes in mid nineties. The average fish production in the last 5 years is about 5.65 lakh tonnes with the contribution of Marine sector being 66 % and Inland sector

34 % for the year. The fish production from the state contributed about 5.86% of India's total fish production for the year 2013-14.

In order to achieve the target of the department of fishery, there is lot of demand for skilled fishery graduates in the state. The data related to the staff positions in the fisheries department have been collected (Sanctioned filled and vacant positions) and estimates have been made on that basis. The estimate showed that there is a 27.38 % vacancy in the state. The number of fishery officers in the state has been estimated at 194 (Table 9.3).

Detailed projected stock demand in department of fisheries is shown in Annexure-4.13.1. The estimated stock requirement in 2014-15 is about 72. This would increase to 117 by 2024-25. Thus, department of fisheries would need about 7 per annum during 2015 to 2025.

9.6.1.2 State Fishery College/ Veterinary/ Agriculture Universities

Different agricultural/veterinary universities in the state employ fishery post-graduates/doctoral for distinct posts from concerned field of their specialization. The SAUs (including all teaching, research and extension) had a combined sanctioned strength of 81 faculties, of which 50 positions (61.73 %) were vacant as on March 31, 2015 details was given in (Table 9.3).

Detailed projected stock demand in farm universities is shown in Annexure-4.13.2. These projections estimate that the stock requirement in 2014-15 is about 50. This would increase to 81 by 2024-25. Thus, SAU's including Teaching, Research and Education; need about 3 per annum during 2015 to 2025.

Table 9.3: Staff positions in the fisheries department and farm universities as on March 31, 2015.

Organisation	Sanctioned	Filled	Vacant	Vacancies %
Universities	81	31	50	61.73
Department	263	194	72	27.38
Total	344	225	122	44.55
Universities - Agriculture and Veterinary				
Department: Fisheries Department, Govt. of Karnataka				

9.6.1.3 Financial Institutions

Establishment Survey showed that about 4 per cent of the total employment (or, 10% of the officers in banks, as officer strength is 40 per cent of the total employment) in banks comprised of graduates and post-graduates in agriculture and allied sciences. About 89 per cent of them were graduates and 3 per cent post-graduate, doctorates forming a mere 0.2 per cent, the rest being diploma/certificate holders in agri-related fields. Employment of officers in public sector banks has been projected to grow at a rate of 2.5 per cent per annum.

Data on number of the agricultural officers employed by the banking institutions are from fishery science qualified is not known as the banking institutions, do not generally prescribe specific qualifications in fishries at the time of recruitment, it may be assumed that the recruitment would be in proportion of the fishery graduates, post-graduates to the total annual intake of graduates in all agricultural fields (which is about 5 per cent, now and may go upto 10% by 2025) as per reports of (Rama Rao *et al.*, 2010) in India and discussion with experts.

Detailed stock projections for financial institutions are given in Annexure-4.13.8. The stock requirement of human capital would grow from 5 in 2014-15 to about 13 by 2024-25. Thus, the requirement is about 2 per annum.

In the state, there is requirement of fisheries graduates in boards and corporations. For instance, in Karnataka over 34 per cent of the officers posts are vacant Table 9.4 and Annexure-4.13.3.

Table 9.4: Staff positions in the boards and corporation as on March 31 2015

Organisation	Sanctioned	Filled	Vacant	Vacancies %
Boards and Corporation	50	33	17	34.00

9.6.2 Private Sector Employment

Fisheries is an important vibrant sector witnessed as the primary source of protein for millions of people. In broad terms, it including inland and sea, aquaculture, gears, navigation, oceanography, aquarium management, fish breeding, processing, export and import of seafood, special products and by-products, research and related activities. Career opportunities apart from for public sector, there are variety of private companies, including agencies and industry/ organizations recruit positions like aquaculture farmer, shellfish culturist, hatchery technician, biological science technician, fish research assistant, *etc*. Many career options exist in private sectors in aquaculture to sea farming of fish, shellfish and marine organisms. Entry-level aquaculture jobs require either a diploma or an under-graduate degree in fisheries, but more advanced positions require a master's or doctorate degree. In private sector, a post-graduate in fisheries sciences has lots of demand to work as Quality Control Officer, Fish Processor, Aquaculturist, Farm Assistant/ Manager, *etc*. The pay varies according to the type of job and specialization of the candidate.

9.6.2.1 Fish Processing

Fish preservation is a very important aspect of the fisheries. Normally the fish farms or other fish capturing sites are located far off from the market place and there is chance of fish decomposition and the uncertainties of their sale in market. When the fishes are caught in numbers, greater than the amount of consumption, their preservation becomes a necessity for their future use. Preservation and processing, therefore, become a very important part of commercial fisheries. It is done in such a manner that the fishes remain fresh for a long time, with a minimum loss of flavour, taste, odour, nutritive value and the digestibility of their flesh. On in average 15 graduates are working in fish processing sector in the state (**Table 9.5**).

Table 9.5: Projected Fisheries Human resource requirement in Fish Processing industries till 2024-25

Year	UG	PG	Ph.D.	Total
2014-15	6	8	1	15
2024-25	8	11	1	20

9.6.2.2 Fish Seed Hatcheries

Fish hatcheries can be important to the survival of freshwater fish species to offset the effects of overfishing and environmental issues affecting fish populations in lakes, rivers and ponds. Determining effective management strategies for fish hatcheries can be a difficult and time consuming task. High quality fish seed is vital in sustaining healthy growth of fisheries. The data on the requirement of technical personnel Fish Seed Hatcheries are presented in Table 9.6 and Annexure 4.13.5.

As per report of NAARM, a fish seed hatchery of capacity 10 to 50 million fry would need a breeder, a hatchery manager, a chemist and an aquaculture engineer (all assumed to be of graduate level), 12 technicians (assumed to be of diploma level) and 10 skilled farmers (assumed to be of certificate level or trained). Considering that the average hatchery is of 160 million fry capacity, these requirements of graduates and diploma holders have been doubled while those of skilled workers have been tripled keeping in view the capacity and economy of scale of hiring the personnel by the industries, and higher norms of 8 graduates, 24 diploma holders and 30 skilled farmers have been adopted. Hence, in Karnataka 4 major seed hatcheries are there, altogether 15 graduates and above are required in hatcheries (**Table 9.6**).

Table 9.6: Projected Fisheries Human resource requirement in fish seed hatcheries industries till 2024-25

Year	UG	PG	Ph.D.	Total
2014-15	8	6	1	15
2024-25	13	10	1	24

9.6.2.3 Fish Feed Industry

Manufactured feeds are an important part of modern commercial aquaculture, providing the balanced nutrition needed by farmed fish. The feeds, in the form of granules or pellets, provide the nutrition in a stable and concentrated form, enabling the fish to feed efficiently and grow to their full potential. Adequate availability of quality fish feed, and its balanced application, is essential for the healthy growth of aquaculture industry. There are about 3 to 4 major fish feed industrial units are there in the state. According to the information collected from establishment survey and experts opinion, about 10 fishery professionals are presently working in fish feed industry in the state and would require an additional 25 by 2024-25 (**Table 9.7**).

Table 9.7: Projected Fisheries Human resource requirement in fish feed industries till 2024-25

Year	UG	PG	Ph.D.	Total
2014-15	5	6	1	10
2024-25	5	6	1	10

9.6.2.4 Other Sectors

There are several other sectors where fisheries science personnel find employment or self- employment to a limited extent. Information collected by establishment survey showed that the extent of movement of fisheries graduates to other sector is comparatively less due to huge demand of graduates in the fisheries filed itself. **In spite of this, the results of establishment survey revealed that quite a few post-graduates and graduates of fisheries science find employment overseas, particularly in the east African and Gulf countries. The demand from all such sectors, such as civil services, fishing equipment industry NGOs, agri-clinics and other self-employment, and global demand has been taken as 3 per cent of the total demand assessed for the sectors discussed above.**

9.7 Summary of Demand Projections

The projections of stocks of horticulture human resources required to meet the needs of various segments of the economy and the consequential flows required from the educational system are summarized below:

9.7.1 Stock Projections

The overall stocks of fisheries human resources required to meet the requirements of various segments of fisheries have been added to give projections upto 2024-25. Three sets of projections have been worked out as explained in the segment- level projections – one a high projection, a low projection and an average. These projections are given in Annexure-4.13. The corresponding annual outputs required from the educational system to lead to these stocks are given in Annexure-4.13.13.

Considering the current supply and experts opinions on graduates aspirations and employee organizations capacity to absorb graduates, the average variant projection is considered desirable. The overall stock as per the average scenario is given in **Table-9.8.** The stock of graduates and above would grow from about 43 in 2014-15 to about 304 in 2024-25, *i.e.,* the average addition to the stock during the period 2015 to 2025 is about 25 per annum.

Table 9.8: Overall Projections of Fisheries Human Resources

Year	UG	PG	Ph.D.	All UG & above
2014-15	22	15	6	43
2024-25	146	98	59	304

9.7.2 Demand – Supply Gap in Stock

In 2014-15, the actual stock from supply is 43. Assuming out turn levels will grow at the same growth rate as observed during the last five years (*i.e.*, 15.69 per cent during 2010-11 to 2014-15), the estimated demand-supply scenario in 2025 will be:

Supply: 189

Demand: 304

Gap: 115 (37.82 % of the demand)

9.7.3 Flow Projections

The overall annual flow required in all sub-sectors of fisheries is given in Annexure-4.1.11 and summary in Fig. 9.1.

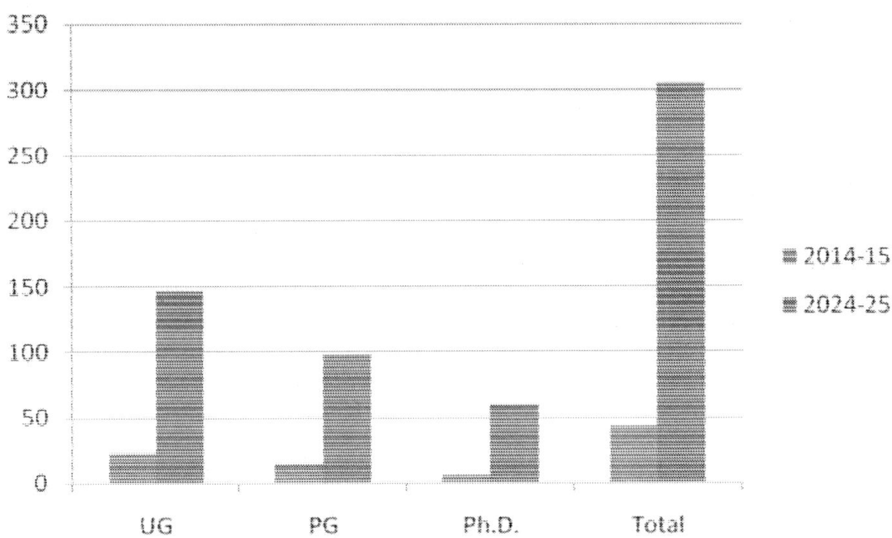

Fig. 9.1: Current (2014-15) and Projected Out turn Required (2024-25) in Fisheries

The projections suggest that by 2024-25, the annual out turn from the fishery education system should be about 146 under-graduates, 98 post-graduates and 59 doctorates, rising from 22 under-graduates, 15 post-graduates and 6 doctorates in 2014-15.

9.8 Conclusions

Looking into the demand for fisheries graduates in the state, there is a need to establish a one more fishery college in interior of Karnataka for development of inland fisheries. Diploma and certification courses in fisheries sciences are to be started to meet the base level extension workers. There is urgent need for filling up of vacancies in teaching faculties in the existing colleges to improve the quality of the fishery education.

Current annual out turn of about only 43 graduates is less than the requirement of 304 by 2024-25. There is need to expand fisheries education to produce additionally 198 graduates (14 UG; 10 PG and 2 PhD) per year by 2024-25.

❑❑❑

Chapter **10**

Supply & Demand Projections, Demand-Supply Gap

This chapter assesses the supply scenario and the demand-supply gap of agriculture and allied sectors human resources at aggregate level.

10.1 Estimation of Supply

Assessment of the available supply of human resource of a particular category was done by cumulating the institutional out turn of the human resources over the years.

The base year (2014-15) supply of agricultural graduates was estimated in the state by indirectly through cumulating of institutional out turns (from all farm universities of the state). Assuming that the average entry of agricultural graduates into the labour force is at about the age of 22 years and exit at the age of 60 years, the working span of agricultural graduates comes to 38 years. In the case of post-graduates and doctorates, the entry age may be taken as 24 and 27, respectively, implying a working span of 36 and 33 years, respectively. Since graduates form about two-thirds of the total out turn, it has been assumed that the average active life-span of agricultural scientists is 37 years. Annual attrition due to mortality, migration and non-participation in work is taken as 1 per cent.

For the period 2010-11 onwards till 2014-15, data on intake and out turn for various disciplines have been collected under the present project from various agricultural universities and colleges in the state. These data include not only graduates but also post-graduates and doctorates as well, who passed out in different years.

Table 10.1: Estimated Outturn of Graduates and Above in Agriculture and Allied Sciences (2010 to 2015)

Degree Programe	2010-11	2011-12	2012-13	2013-14	2014-15
Agriculture	1160	1102	1214	1416	1718
Agriculture Marketing	77	92	106	141	105
Agriculture Engineering	89	79	88	123	98
Agriculture Biotechnology	75	72	82	81	91
Rural Home Science	58	56	51	62	73
Food Technology	36	39	50	50	46
Sericulture	18	24	18	28	31
Forestry	55	73	133	110	99
Horticulture	275	307	411	508	404
Veterinary	163	241	219	244	246
Fishery	28	28	31	51	43
Dairy Technology	35	50	49	54	47
Total	**2069**	**2163**	**2452**	**2868**	**3001**

10.2 Estimates of Supply and Demand in the Base Year, 2014-15

Based on the data collected from various colleges and farm universities in the state are cumulated to estimate the supply in the base year. These are given in the Table 10.2.

Table 10.2: Estimates of Supply of Stock of Agricultural Graduates and Demand in the Base Year, 2014-15

Discipline	Estimated		Demand-Supply gap	
	Demand in 2014-15	Supply in 2014-15	Number	As % of demand
Agriculture	3931	1718	2213	56.3
Agriculture Marketing	415	105	310	74.7
Agriculture Engineering	170	98	72	42.4
Agriculture Biotechnology	108	91	17	15.7
Rural Home Science	96	73	23	24.0
Food Technology	76	75	1	1.3
Sericulture	391	31	360	92.1
Forestry	1308	99	1210	92.5
Horticulture	1274	404	870	68.3
Veterinary	1486	246	1240	83.4
Fishery	195	43	152	78.0
Dairy Technology	191	47	144	75.4
Total	**9641**	**3030**	**6612**	**68.6**

Thus, the total stock of agricultural graduates and above in 2014-15 has been estimated at about 3030, falling short of the assessed demand of 6612 by 68.6 per cent. The shortfall is across the board among all disciplines barring Forestry, Sericulture and Veterinary.

10.3 Projected Supply till 2024-25

The estimates of supply have been projected till 2024-25 under varying sets of assumptions about the expansion of agricultural education. These assumptions are:

Scenario A: The average out turn levels of the last five years (2010-11 to 2014-15) will be maintained unchanged till 2024-25.

Scenario B: The out turn levels will grow at the same rate as the growth observed during the last five years beginning with the average out turns during that period, given above.

Scenario C: The out turn will expand by 10 per cent per year till 2024-25, beginning with the average level of last five years in 2014-15.

10.4 Demand - Supply Gap by 2024-25

The projected demand for various categories of human resource is compared in the following (Table-10.2 and 10.3), with the likely supply situation under various scenarios described above to identify the demand-supply gaps.

Table 10.3: Estimates of Supply of Stock of Agricultural Graduates and Demand for 2024-25

Discipline	Demand in 2024-25	Supply under scenario			Demand-supply gap (% of demand)		
		A	B	C	A	B	C
Agriculture	6495	1718	4904	6667	74	24	-3
Agriculture Marketing	681	105	305	691	85	55	-2
Agril. Engineering	355	98	186	365	72	48	-3
Agril. Biotechnology	282	91	152	291	68	46	-3
Rural Home Science	240	73	128	248	70	47	-3
Food Technology	416	75	233	423	82	44	-2
Sericulture	464	31	104	467	93	78	-1
Forestry	1463	99	497	1472	93	66	-1
Horticulture	2118	404	1443	2158	81	32	-2
Veterinary	1697	246	551	1722	86	68	-1
Fishery	304	43	189	308	86	38	-1
Dairy Technology	316	47	120	320	85	62	-1
Total	**14830**	**3030**	**8813**	**15133**	80	41	-2

It may be observed that under in a and b scenarios, there is likely to be substantial gap between demand and supply of human resources, but in scenario c there is a excess supply in all the sectors.

10.5 Pattern of Agricultural Human Resources Deployment

The deployment of agricultural human resources during 2014-15 is depicted Fig-10.1 (all disciplines put together) and 10.2 (discipline-wise). The shares of various segments by employment are: 81 per cent in government, 9.04 per cent in private, 4.37 per cent in financial, 1.00 NGO's, 0.14 per cent self-employment and 4.45 per cent in others sectors. The major shift will take place in the coming decades (2024-25) may be due to decline in the share of public sector in employment, and also due to freezing employment in government sector as well as expansion of opportunities in the private sector. This is in tune with the emergence of commercialization as well as diversification.

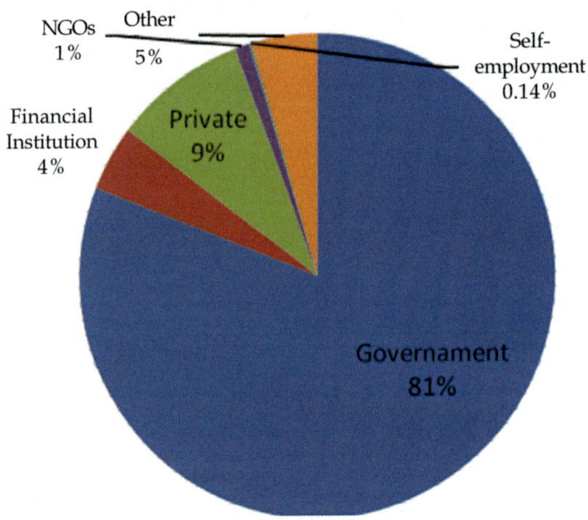

Fig. 10.1: Pattern of Agricultural Human Resources Deployment, 2014-15

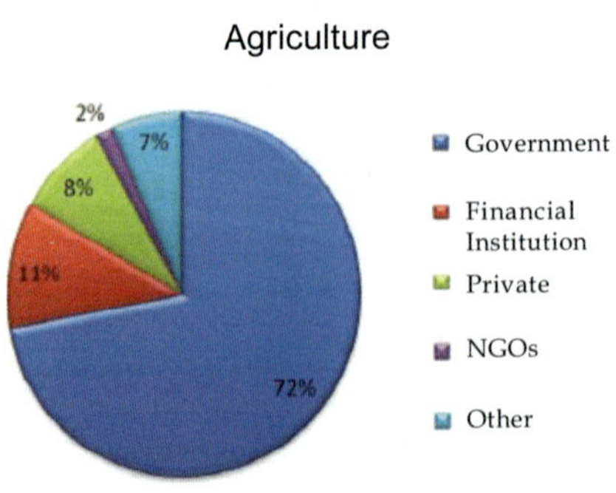

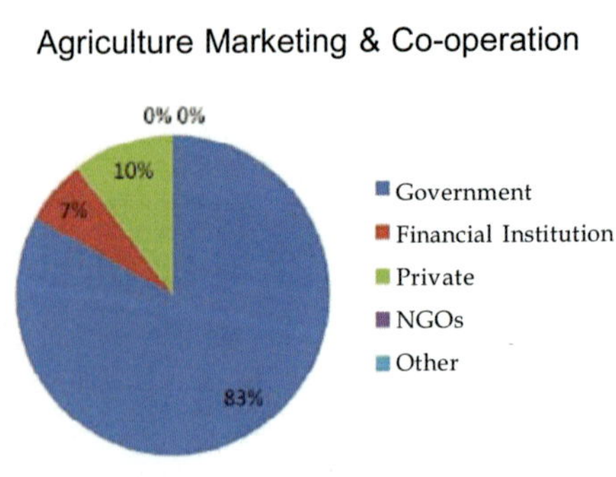

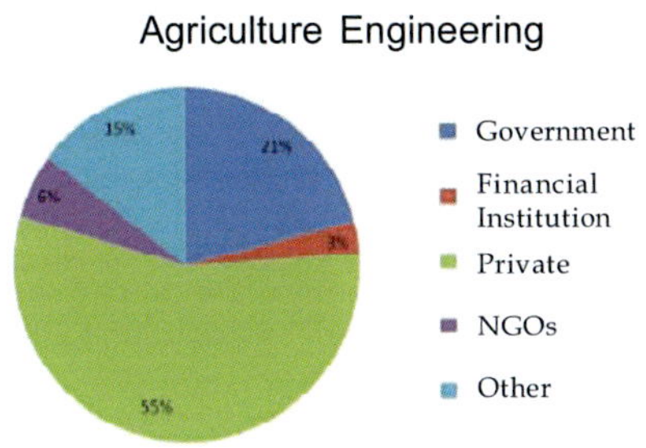
Agriculture Engineering
15%
21%
6%
3%
55%
Government
Financial Institution
Private
NGOs
Other

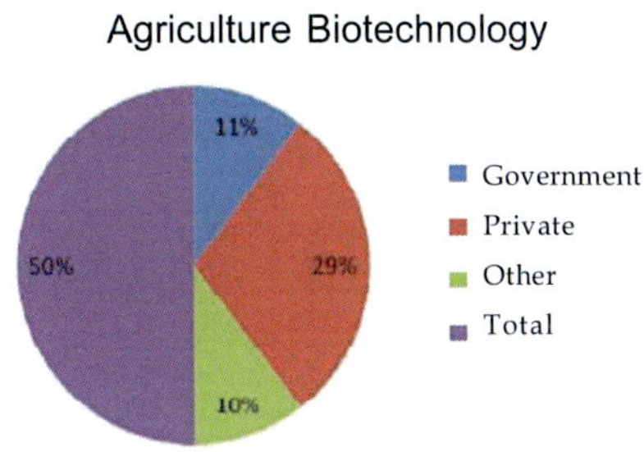
Agriculture Biotechnology
11%
50%
29%
10%
Government
Private
Other
Total

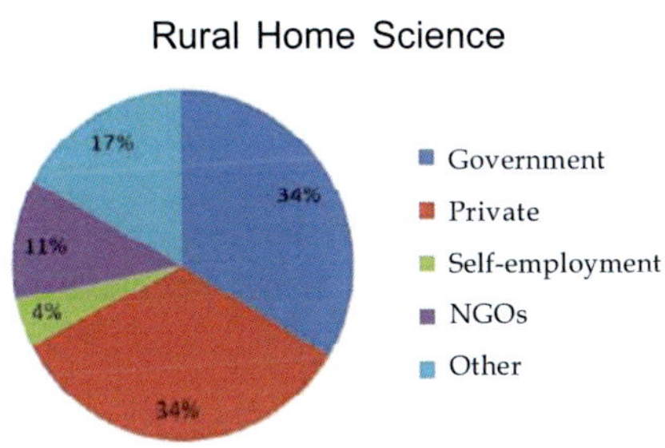
Rural Home Science
17%
34%
11%
4%
34%
Government
Private
Self-employment
NGOs
Other

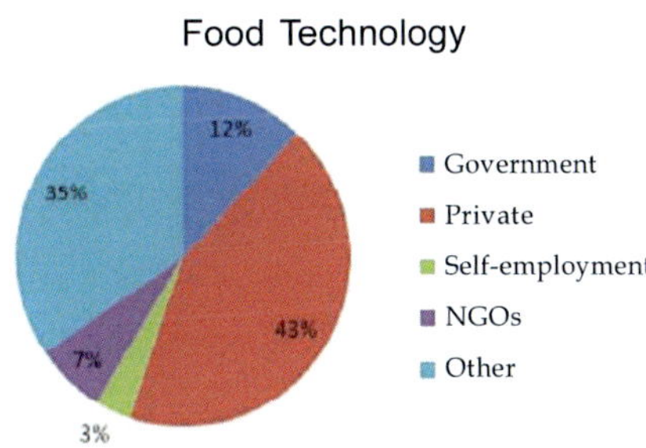
Food Technology
12%
35%
43%
7%
3%
Government
Private
Self-employment
NGOs
Other

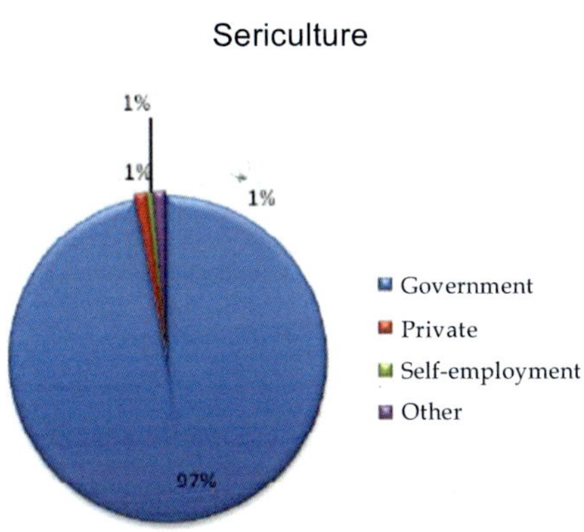
Sericulture
1%
1%
1%
97%
Government
Private
Self-employment
Other

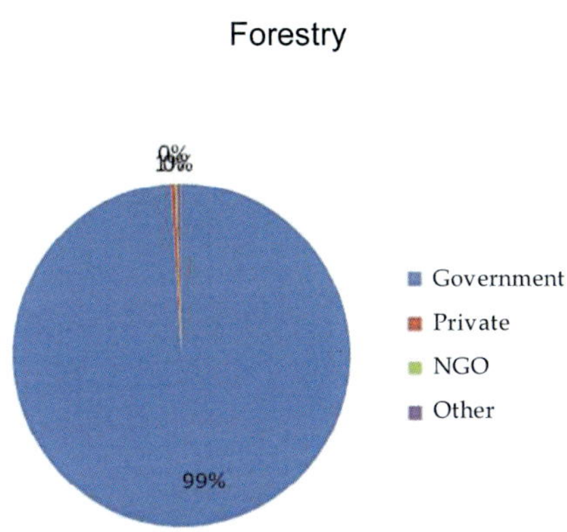
Forestry
99%
Government
Private
NGO
Other

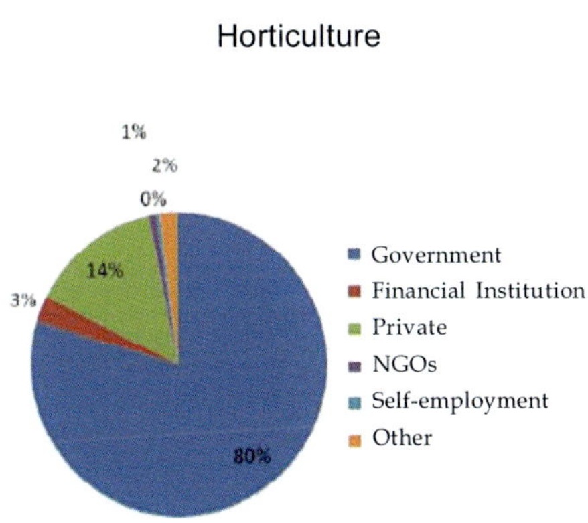
Horticulture
1%
2%
0%
14%
3%
80%
Government
Financial Institution
Private
NGOs
Self-employment
Other

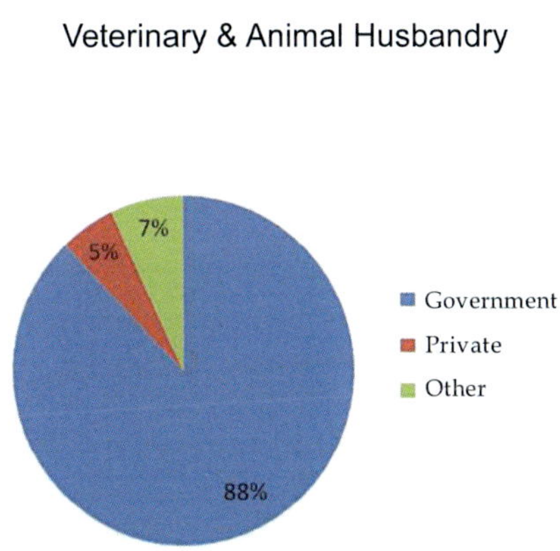
Veterinary & Animal Husbandry
7%
5%
88%
Government
Private
Other

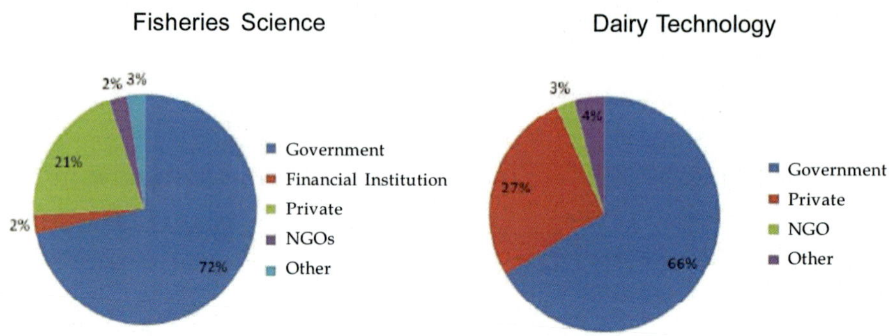

Fig. 10.2: Profile of Deployment of Agriculture Human Resources in Different Sectors

10.6 Annual Flow of Supply-Demand

With the demand projections made in the earlier chapters for various sub-sectors, the required annual out turn for the year 2024-25 in comparison to the same during 2014-15 is given in Table 10.4.

Table 10.4: Required Annual Out turn by Education Level in 2014-15 (Supply) and 2024-25 (Demand)

Discipline	UG		PG		Ph.D.		UG and above	
	2014-15	2024-25	2014-15	2024-25	2014-15	2024-25	2014-15	2024-25
Agriculture	955	3309	637	2014	126	1056	1718	6379
Agril. Marketing	69	456	33	221	3	61	105	738
Agril. Engineering	74	151	23	160	1	45	98	355
Agril. Biotechnology	46	110	36	104	9	59	91	272
Rural Home Science	39	100	24	88	10	50	73	238
Food Technology	75	314	0	69	0	32	75	416
Sericulture	15	293	12	140	4	31	31	464
Forestry	60	955	35	451	4	56	99	1463
Horticulture	249	1074	120	677	35	367	404	2118
Veterinary	151	971	77	502	18	224	246	1697
Fishery	22	146	15	98	6	59	43	304
Dairy Technology	38	189	7	98	2	29	47	316
Total	**1793**	**8067**	**1019**	**4624**	**218**	**2068**	**3030**	**14760**

At present, the existing education system is producing about 3,030 graduates per year with crop sciences contributing 2/3rd of it. The projections indicate that by 2024-25, the annual out turn required would have to be about 14,760. Based on the current supply, the demand supply gap would be 11,730 (Fig. 10.3).

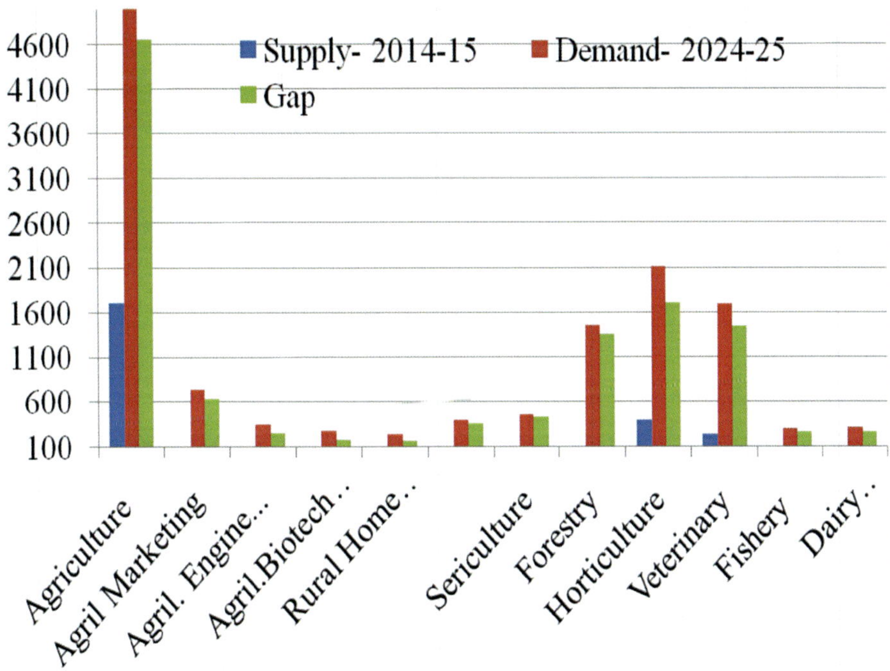

Fig. 10.3: Discipline-wise Net Additional Out turn Required by 2024-25

Discipline-wise, the additional annual requirements of out turns are expected to be: Agriculture-4661; Agriculture Markrting-633; Agriculture Engineering-257; Agriculture Biotechnology-181; Rural Home Science-165; Food Technology-341; Sericulture-433; Forestry-1364; Horticulture-1714; Veterinary-1451; Fishery-261 and Dairy Technology-269. The average out turn during the last decade grew at the rate of five per cent per annum and in the coming decade, it is projected to grow at above five per cent to meet the demands.

10.7 Degree-wise Requirement by 2024-25

The required annual increase in out turn by 2024-25, by education level and discipline is given in Table-10.5 The current level of graduate, post-graduate and Ph.D. output need to be increased by 2.3, 1.9 and 3.0 times, respectively.

Table 10.5: Required Annual Increase in Out turn of Graduates by 2024-25

Discipline	UG	PG	Ph.D.	UG and above
Agriculture	2354	1377	930	4661
Agriculture Marketing	387	188	58	633
Agriculture Engineering	77	137	44	257
Agriculture Biotechnology	64	68	50	181
Rural Home Science	61	64	40	165
Food Technology	239	69	32	341
Sericulture	278	128	27	433
Forestry	895	416	53	1364
Horticulture	825	557	332	1714
Veterinary	820	425	206	1451
Fishery	124	83	53	261
Dairy Technology	151	91	27	269
Total	**6274**	**3605**	**1851**	**11730**

10.8 Additional Colleges to be Established

At the current rate of out turn is less than 65 per cent of the intake capacity. Though full capacity utilization is not possible, the system efficiency can be improved to produce other 15-20 per cent additional graduates from the existing system. The expansion may have to consider this factor.

The demand of professionals at PG and Ph.D. level would be met through increase in number of sanctioned seats in existing colleges. At UG level, considering the current intake capacity and large gap in demand-supply, there is a need to establish about 20 new colleges. Discipline-wise new colleges needed are given in Table 10.6.

Table 10.6: Additional Colleges Needed at UG Level

Discipline	No. of Colleges
Agriculture	3
Agriculture Marketing	1
Agriculture Engineering	1
Agriculture Biotechnology	1
Rural Home Science	1
Food Technology	1
Sericulture	1
Forestry	1
Horticulture	1
Veterinary	2
Fishery	1
Dairy Technology	1
Total	**15**

The new colleges would also require adequate infrastructure and qualified teachers to ensure quality of education. To meet the projected demand and required out turn growth and various issues pertaining to education quality and management important observations and conclusions are given below under various components:

10.9 Conclusions

Compared to other streams of science and technology, the growth in agricultural education is relatively low (3 per cent) during the period, 1970 -2000 in India (NAARM report). In Karnataka, looking into the past five years (2014-15) data, the growth has accelerated to >5%, the result of which can be seen from the increased out turn of students in the last five years specially with the increase in the intake in the existing colleges and emergence of new colleges in agricultural education. Considering the current developments and projected growth of various sectors of agriculture in the next decade, agricultural education is likely to experience above 5 per cent growth per annum. On this basis, the scenario B presented is most likely. On this scenario, there will be a gap of 41 per cent in demand-supply.

❑❑❑

Chapter **11**

Review of ICAR and SAU Schemes for HRD of Students

11.1 Introduction

There is increasing evidence and recognition that human resource, is the capability of people to be effective and productive economic agents matters more for development than natural resources and man-made physical capital. In the particular case of agriculture, most studies on the subject establish that the education and skills of agricultural people are significant factors in explaining the inter-farm and inter-country differences in agricultural performance, along with the more conventional factors such as availabilities of land and water resources, inputs, credit, *etc.*

ICAR has planned and implemented several schemes and programs for strengthening and development of agricultural education through Centres of Advanced Studies, Niche area of excellence, University level-textbook writing, National Fellows and Professional Chairs, infrastructure upgradation, faculty competence building, accreditation, incentives through best teacher award, scholarships to students and their practical training in laboratory and real life field situations. Entering into MoU with the other universities in the world for co-operation in agricultural education and research by the ICAR and SAUs for agricultural education are the new initiatives, which have immense potential for enriching agricultural education as well as to reach the unreached.

With the shrinking of per caput agricultural resources following demographic growth, with the agricultural labour force in the developing countries projected to continue at positive (though declining) growth rates and with the share of young people in the total also continuing to grow, the task of upgrading the literacy, the skills and other capabilities of the agricultural people is enormous, for coping with both the increases in numbers and the backlog inherited from the past. Moreover, the increasingly binding character of natural resource scarcities imposes severe limits on the extent, to which production increases can be had through expansion of extensive agriculture. The generation and diffusion of technology and management capabilities for more intensive and modernized agriculture and supporting services become imperative. This can only be achieved through the upgrading of the quality of human resources employed in agriculture.

It is noted that many dimensions of the human resources development (HRD) issue are final end-objectives of development, e.g. boost up of skill set and technical knowledge, *etc*. Although this chapter is concerned with policies to upgrade the quality of graduates, by reviewing of ICAR and SAU schemes for HRD of students in Karnataka. Hence, need to make progress in skill development, through various schemes of ICAR and state government, through farm universities. This is important, since it implies that evaluation of basic education system versus creation of more directly productive agricultural skills.

11.2 Scholarships

Scholarships are award of financial aid to enable students in their academic pursuit and/or further their education. Scholarships are awarded based on various criteria, which usually reflect the values and purposes of the donor or funder of the award. Scholarship money is not required to be repaid, as it is not loan.

There are different types of scholarships for the agriculture and allied sector students at graduate, post-graduate and doctoral level.

11.2.1 International Scholarships

11.2.1.1 Indo-African Fellowship

Department of Science and Technology (DST) and Ministry of External Affairs (MEA), Government of India (GoI), through the Federation of Indian Chambers of Commerce & Industry (FICCI) have launched the C.V. Raman Fellowship for African Researchers programme under the India-Africa Forum Summit to promote human capacity building through scientific and technological co-operation between Africa and India. The objective of this Fellowship is to provide opportunity for African researchers to conduct collaborative research in various areas of science and technology in different Indian universities and R&D institutions under guidance of host scientists, in India. This prestigious fellowship is aimed to further strengthen the bond between India and African nations, in the fields of science and technology. Federation of Indian Chamber of Commerce & Industry (FICCI) is the 'Coordinating Partner' for C V Raman International Fellowship Programme. FICCI, on behalf of GoI, invites applications from African researchers for award of this fellowship for conducting research in Indian institutions across all areas of science and technology.

11.2.1.2 Monsanto Scholarship

The Monsanto Graduate Student Scholarship is the part of the company's broader focus on innovation and investment in agriculture. To meet the challenges, the world faces in feeding nine billion people by 2050, investing in future leaders pursuing STEM careers in food and agriculture is key. The next generation of innovators will be the ones to ascend and meet the challenges of global food security. The establishment of Monsanto's STEM Fellowship is part of the company's investment in agriculture to help increase global food production in the face of growing demands, limited natural resources and a changing climate. Monsanto has made a $1 million commitment over three years to support high achieving minority students pursuing a Ph.D. degree in agriculture or STEM related discipline. The Monsanto Fund 1890s Student Scholarship is a part of the company's broader focus on innovation and investment in agriculture.

11.2.1.3 Monsanto's Beachell-Borlaug International Scholars Program

Those students working in the rice and wheat crops, get the benefit of Monsanto's Beachell-Borlaug International Scholars Program. Because these two crops are the staple food crops and critical to the food security of billions of people around the world. Together, they feed more than half of the world's population. However, yield improvements in rice and wheat lag behind other crops. As a part of this monsanto's commitment to sustainable agriculture, Monsanto Company has pledged a total of $13 million to provide fellowship opportunities to highly motivated individuals seeking their Ph.D. in rice or wheat plant breeding. Hence, these fellowships, the scholars are very much helpful to gain the leadership skills, education and tools to positively impact rice and wheat farmers around the world and improve the future of agriculture.

11.2.1.4 International Crops Research Institute for the Semi-Arid Tropics (ICRISAT)

This is an international organisation, which conducts agricultural research for rural development, headquartered in Patancheru (Hyderabad, Telangana, India) with several regional centers (Bamako (Mali), Nairobi (Kenya)) and research stations. There are various scholarships, available for Indian students pursuing agriculture education at state Agriculture Universities and ICAR research Institutions at post-graduation and doctoral level.

11.2.2 National Scholarships

11.2.2.1 National Talent Scholarships (NTS) for Under-graduates (ICAR)

The National Talent Scholarships is the ICAR scholarship scheme for the students of under-graduates, studying in Agricultural Universities (B.Sc./B.Tech./B.F.Sc., *etc.*) not located in the domicile State of the student and selected through ICAR/VCI All India Entrance Examination for admission to UG programs in the month of May every year and award of NTS in agriculture and allied subjects *viz.*, Agriculture, Horticulture, Agricultural Engineering, Dairy Technology, Forestry, Food Science, Fishery Sciences, Home Sciences, Sericulture and Agri Marketing Banking & Co-operation. Scholarship funded by **Indian Council of Agricultural Research (ICAR)**. Amount of fellowship is Rs. 1000/- pm for the duration of the program.

11.2.2.2 DBT-HRD Scholarships

DBT biology scholarship is granted to 100 students from schools around the country. The students must be of higher secondary, intermediate or 10+2 level from any of the boards—CBSE, ICSE or state boards. This is done to encourage students to pursue studies in biological sciences and related subjects after 10+2. A total of 44 students from CBSE board, four from ICSE board and 52 (two from each state board) from the various state boards are selected for the award, which carries a cash prize of Rs. 20,000, a certificate and a medal.

11.2.2.3 DBT- JRF

The department provides fellowships to biotechnology students to pursue doctoral research in universities/research institutions across the country. The students are selected through Biotechnology Eligibility

Test (BET) now being co-ordinated by the National Centre for Cell Sciences.

11.2.2.4 DBT- SRF/RA/ Associate ship

The department provides fellowships for post-doctoral research in frontier areas of biotechnology and life sciences at premier institutions in India. The DBT-RA programme is being implemented by the Indian Institute of Science, Bangalore. The fellowship is initially awarded for a period of two years and the support can be extended for 1 to 2 years based on review of progress.

11.2.2.5 Sir C.V. Raman Scholarship

In order to increase the number of students, studying in science courses and to encourage them to take up science courses from 2001-02 onwards for students in government and private aided degree colleges, this scholarship is awarded. Amount of scholarship Rs.5000/- per year. Fresh awards 200, Renewals depending upon the no. of applications received, Amount Sanctioned, *etc.*

11.2.2.6 National Merit Scholarship

This is being awarded to students to provide financial assistance to the meritorious students of rural areas, studying in degree and post-graduate level in colleges and universities. Eligibility: Students should have obtained 60% in science stream. Scholarships will be paid only to those students whose parents'/guardian's income from all sources does not exceed Rs.1.00 lakh per annum. Amount of Scholarship: Rs. 500 per month for degree students and Rs. 750 for PG students.

11.2.2.7 Govt. of India Scholarship for SC/ST Candidates

The only SC/ST candidates are eligible for the award of these scholarships. The parents' income limit should not exceed Rs. 1.00 lakhs per annum. Students, who avail the benefit of G.O.I. scholarship in one course, are eligible to claim the same benefit for higher studies. Students, who are employed, are not eligible. In a family only two male children are eligible for the scholarship, but there is no restriction for female children. The students, who are eligible to avail the G.O.I. scholarship, are not eligible to avail of any other scholarship. Amount of Scholarship, Rs. 185.00 for day scholars and Rs.335=00 for hostellers per month. Similarly, state government is also having some of the scholarship schemes for the SC/ST students.

11.2.3 ICAR - Hands On & Implant Training (RAWE)

The scheme envisages the introduction of Rural Awareness Works Experience (RAWE) programmes in the entire universities curriculum as an essential pre-requisite for the award of the degree to ensure adequate practical training. The RAWE will include precisely prescribed programmes of work in farms as well as KVKs instructional plants, *etc.* for a period of one semester with provision for adequate supervision, reporting and examination at the end. The programme shall be introduced for one semester under the structured degree programmes. A student, undergoing RAWE, will not be allowed to avail of any other fellowship/scholarship during tenure of stipend of the Council. In case a candidate is already receiving any other fellowship/Scholarship, it will be surrendered by him before accepting stipend of the Council. Merit cum means scholarship, Free ship is, however, not covered under the above conditions.

11.2.4 State Government Scholarships

11.2.4.1 Government of Karnataka

The award of State Post-Matric scholarship was introduced to enable the students to continue the studies, where he/she is not eligible for award of G.O.I. Scholarship. Value of scholarship Rs. 400/- per annum. Eligibility is she/he should belong to SC category. Parent's income should not exceed Rs. 12,000.00 per annum. Students who fail once in degree courses are eligible for state Post-Matric scholarship. Even 3rd or 4th Male children of the family are eligible for state Post-Matric scholarship.

11.2.4.2 Handicap Scholarship

The state of Karnataka provides educational scholarship for 10 months in each academic year. Amount is Rs. 250/- per month for graduates and Rs. 300/-per month for post-graduates. Eligibility criteria includes: Passing marks in previous annual examination, above 40 disability persons, need not give family income certificate, there is exemption from maximum income limit, and Resident of Karnataka state is must.

11.2.4.3 Karnataka e-pass

Karnataka e-pass is one of the important welfare measures of the Government of Karnataka, which is aimed at the educational upliftment of Students, initiated by the Department of Backward Class Welfare. The eligible students can apply Online to get post-matric scholarships. The system links all welfare departments, treasury, databases of Secondary Schools Leaving Certificate (SSLC), Colleges and Banks to disburse scholarships. There are various scholarships coming under this link *viz.*, State Govt. Merit Fellowship, Students Aid Fund Scholarship, Fee Concession, BCM Scholarship, Karnataka Minorities Scholarship, EBL from BCM scholarship, Vidyashree, *etc.*

11.2.4.4 Promotion of University Research and Scientific Excellence (PURSE) Scholarship

In appreciation of the R&D contributions of the performing universities, PURSE initiative is taken by Department of Science & Technology, Government of India to initiate value added proactive measures through introduction of R&D Incentive Grant. PURSE scheme provides substantive research grant to universities based on scientific publications in Science Citation Indexed Journals, through SCOPUS Database. Hence, such grants can be used for the scholarship purpose to the post-graduates and doctoral in the universities. Amount varies from college to college and availability of fund is on an average, the PG students are getting Rs. 5000 and Ph.D. students gets Rs. 10,000.

11.2.5 University Scholarships

All the farm university are having scholarship schemes for meritorious students. The scholarship can be given to the students based upon the performance of the marks in the previous exam. Number of scholarship depends upon the total number of students, for every 10 students, one merit scholarship will be given. The amount of scholarship ranges from Rs. 1000 to 2000 pm (UG-1200, PG-1500 and Ph.D-2000). Similarly, the freeship for economically poor and meritorious post-graduates and doctoral students.

11.2.6 Other Scholarships

11.2.6.1 Jindal Scholarship

Jindal Aluminium Ltd, Jindal Nagar, Tumkur Road, Bangalore-560073 and Sitaram Jindal Foundation, 11 Green Avenue, Vasant Kunj, New Delhi–110070, which are consistently engaged in the field of Charitable and philanthropic activities in India since 1969, in furtherance to their said endeavour and sublime approach for the benefit of the general masses, have instituted the 'Sitaram Jindal Research Fellowship Scheme' to encourage and inspire the talent and recognize excellence in innovation and research. The number of scholarship and amount varies from degree programme to degree programme and one academic year to the other.

Apart from above mentioned scholarships and schemes, some other scholarships *viz.*, research fellowship funded by private or public agencies/ companies/ institutions and other national and international scholarships are also there for students, studying in the farm universities.

11.2.7 Student's Opinion about the ICAR, State Government and SAUs Scholarships

Scholarships awarded are based upon various criteria, which usually reflect the values and purposes of the donor or founder of the award. Majorly, scholarships are to be awarded to people who are meritorious, poor and deserving. The results of the survey and discussion with students, teachers and education experts indicated that in the farm universities, various scholarships and schemes for the students are available. There is a need for increased number of awards and quantum of scholarships to students from poor background and not to students from high socio-economic status by the government, Non-governmental organizations, and other philanthropist to alleviate this problem. There is need to increase the number as well as quantum scholarship amount in order to attract the talent and meritorious students to the farm sciences.

11.3 Review of HRD Programmes/Schemes

The students were asked to give their opinion about the usefulness of following programmes, RAWE programme, experiential learning, field visits (Farms/Industries/Institutions), state tour, all India tour and extra-curricular activities.

Results indicated that about 50 per cent students are fully satisfied about above mentioned activities but remaining 30 per cent are somewhat satisfied and 20 per cent are not at all satisfied. From these results, it is clearly indicated that there is need to improve in RAWE programme, experiential learning, field visits, educational tour and extra – curricular activities for the overall improvement in the education system. Because next to the courses taught in a study, also the extra-curricular opportunities like internships, guest lectures or opportunities to attend conferences matter with regard to quality and attractiveness of a study. All the interviewees agreed to the fact that extra-curricular activities are a great extension to the study and that it increases the attractiveness of it a lot. By actively taking part in these types of activities, they say they will get more directly involved with all the interesting things going on in the agro-food sector, increasing your commitment for both, your study and also these external activities. Fortunately, the range of extra-curricular opportunities that is available to agriculture related students are numerous. Not only are they organized by universities or the colleges but also there is an increasing trend of big national level events organized by ICAR and try to get youth actively involved in the extra-curricular activities.

To increase the number of students that choose to study agriculture or related sciences, several actions can be taken. Firstly, realize the importance of agriculture for national economy and food security and its sciences, if needed. Making students more aware of the role of agriculture in our daily life's, of the challenges, all of us are facing regarding food security in the future and of the many different aspects, the farm to fork chain contains, is a way to increase awareness of the importance of the sector, improve its images and with it possibly attract more young people. Furthermore, both the career perspectives as well as the extra-curricular activities of the agricultural sector, are needed to be change. At present, average students spend 1.84 h per day, 5.56 h per week, and 12.44 h per month for extra-curricular activity.

The results showed that, there are about 43% of UG, 45% PG and 57 % of Ph.D. students are presently getting one in other scholarship. The amount varies from degree programme to programme; on in average Rs. 1000 for UG, 2000 for PG and 10,000 for Ph.D. students receiving scholarships.

11.3.1 Conduct and Usefulness of HRD – Students Programmes

The students fully satisfied about conduct and usefulness of different programs *viz.*, RAWE programme (56 and 54 %), field visits including Farms/Industries/Institutions (43 and 51 %), Tours (60 and 50%), extra-curricular activities (40 and 45%), respectively. Teacher also expressed similar opinions with respect to conduct and usefulness of RAWE programme, field visits, tours and extra-curricular activities. The details of HRD of student programmes are discussed in the next chapter (Chapter 12).

Table 11.1: Extent of sample students benefitted by scholarships (UG)

N=184

Sl. No.	Name of Scholarship	Percentage of Students (n=80)	Amount (per Month)
I State Govt.			
1	E-Pass (Vidyashree)	27.50	450
2	OBC Scholarship from Social Welfare Dept.	13.75	500
3	State Govt. Scholarship for General Merit	3.75	300
4	Fee Concession	7.50	700
5	Food and Accommodation charges	2.50	800
6	Minority Scholarship (BCM)	3.75	800
II Govt. of India			
1	NTS	3.75	1200
2	GoI	6.25	1000
III	**ICAR, for UG**	5.00	2000
IV University			
	UAS, Merit	11.25	1200
	General Merit	3.75	1000
	Stay fund	2.50	900
	Incentive	2.50	500
V Other Private			
	Jindal	6.25	800
		100	868
	No. of Students benefitted	80	
	Students Benefitted (%)	**43**	

Table 11.2: Extent of sample students benefitted by scholarships (PG)

N=40

Sl. No.	Name of Scholarship	Percentage of Students (n=18)	Amount (per Month)
I	**State Govt.**		
1	E-Pass (Vidyashree)	16.67	600
2	OBC Scholarship from Social Welfare Dept.	11.11	1000
3	Fee Concession	11.11	900
II	**Govt. of India**		
1	NTS	16.67	1200
2	GoI	11.11	1000
III	**ICAR, JRF**	**11.11**	**8960**
IV	**University**		
1	UAS, Merit	16.67	1500
2	Stay fund	5.56	1200
		100	2045
	% PG Students Benefitted	**45**	

Table 11.3: Extent of sample students benefitted by scholarships (Ph.D.)

N=21

Sl. No.	Name of Scholarship	Percentage of Students (n=12)	Amount (per Month)
I	**State Govt.**		
1	E-Pass (Vidyashree)	33.33	600
2	Fee Concession	16.67	1000
II	**Govt. of India**		
1	INSPIRE	8.33	25000
2	RJNF	8.33	25000
III	**ICAR, SRF**	**16.67**	**16000**
IV	**University (Merit)**	**16.67**	**2000**
		100	10071.4
	% Ph.D. Students Benefitted	**57**	

❑❑❑

Chapter 12

Assessment of Quality of Education in Agriculture and Allied Sectors

12.1 Introduction

As mentioned in the project methodology, execution of the project began with a massive data collection effort to develop a basis for estimation of agricultural human resources. The data collected included both, quantitative and qualitative. In order to assess the quality of education, stakeholder survey has been conducted with a targeted respondents such as students, teachers, agricultural working graduates, establishments, institutions and organizations (**Table 3.2, 3.3, 3.4**).

12.2 Coverage

An effort was made to cover all the sectors (Agriculture and allied sector) and all types during survey. The individual educational experts have been covered from State Agricultural Universities, government departments as well as other establishments in public and private sectors. 50 experts functioning in various disciplines like Agriculture, Horticulture, Veterinary, Agril. Engineering, Agril.Marketing, Bio-Technology, Dairy Technology, Fishery, Forestry, Rural Home Science and Sericulture have been covered.

Teachers survey- designation-wise coverage shows that about 50 per cent of the experts are professor, 30 per cent are Associate Professor and 20 per cent are Assistant Professor, Establishment/Organizations coverage shows that about 60 % public and 40 % private sector organizations. Students-wise 20 per cent had been with doctorate degrees, while 35 per cent were under-graduates, 40 per cent were post-graduates and 5 per cent diploma holders. A good coverage has been given to female students (50 %) of various disciplines. Alumni working in public and private organizations from all the farm universities were covered on random bases. There are about 30 alumni who approached personally and all have responded, which includes under-graduates (35 per cent), post-graduates (45 per cent), doctorates (10 per cent) and 10 per cent diploma holders. These graduates belong to various disciplines of agriculture with fair representation by gender and rural/urban background.

In the establishment survey, coverage of different colleges from each discipline from all the farm universities sector-wise. 30 establishments engaged in agriculture, dairy, food processing, bio-technology, financial institutions, *etc.* have been covered. About 60 of the establishments had been in private sector while the rest are in public sector, co-operative and self-employed. Location-wise, 50 per cent are in rural areas and 50 per cent in urban areas.

12.3 Major Results

12.3.1 Survey of Students in Agriculture and Allied sectors

Mainly due to exponential population growth, world agricultural production has to increase by at least 60 per cent within the next 35 years (Alexandratos, 2012). At the same time, almost all arable land has already been taken into cultivation and vast expansion of agricultural area is no option. This means that one of the main ways to meet this increasing food demand, must come from an increased yield per hectare. To establish this increase in productivity, it is essential that knowledge of agriculture is applied into the fields. Now-a-days a lot of agricultural research is being done to find the best techniques to meet this increasing demand. However, the number of students interested in studying agriculture or any related science is lagging behind, which causes a hampering of the innovation and application of agricultural knowledge and skill.

To find out the skill gap and quality of agricultural education, 245 students who study in the field of agriculture and related sciences have been interviewed through structured questionnaires. These students include (184-UG, 40-PG and 21- Ph.D.) from different colleges in all the farm universities of the state. Questions related to the quality of the education system, in particular about the course curriculum, Teaching, infrastructure facilities, the extra-curricular activities, etc. were covered.

The results revealed that irrespective of the degree programmes, there are about 68 per cent students, joined agricultural degree by choice and remaining are by chance. Among the students who joined degree by choice are opined that, they joined agricultural degree due to best available option are 45, 43 and 33 per cent UG, PG and Ph.D. students, respectively, to gain practical knowledge about agriculture to enable me to practice progressive agriculture are 44, 55 and 29 per cent UG, PG and Ph.D. students, respectively, job prospects are much better in agriculture career, agriculture are 38, 48 and 38 per cent UG, PG and Ph.D. students, respectively, good course for preparing for Civil Service Examinations are 28, 33 and 24 per cent UG, PG and Ph.D. students, respectively, since I did not get any job with my previous qualification are 13, 5 and 10 per cent UG, PG and Ph.D. students, respectively. About 32 per cent graduates are joined degree by chance. In order to reduce this first step is to increase the general interest and awareness to students in the high school/PUC level about agriculture sector (Table 12.1.2).

12.3.2 Course Curriculum

In the interviews, the students were asked their opinion regarding the quality of their Course Curriculum. Since the outcome is, naturally, strongly dependent on degree programme, and sectors, a generalized answer can't be given. They were also asked about suggestions what the changes make, are in the agriculture education system to make improvement in the education. Few of the student comments that the content of the study should be adapted to the burning issues in society and able to solve the farmer's problems, there is need to increase the field visits, industrial exposer tours and training for improvement in the soft skills. If these types of changes were implemented in their course curriculum, to create a direct link between their education and the societal debate on these topics. The results of student's survey; opinion of UG students about present course curriculum 47 % of the UG students strongly agree that needs to include more field visits, 43, 36 and 41 % agreed that matches required in self-entrepreneurship, job in govt. and private, respectively. Master degree students opinioned that, 60 % strongly agreed that needs more field visits 63, 35 and 48 % agreed that matches required in self-entrepreneurship, job in govt. and private, respectively. Whereas, 38 % of the Ph.D. students strongly agree that needs to include more field visits, 52, 38 and 38 % agreed that matches required in self-entrepreneurship, job in govt. and private, respectively (Table 12.1.6 to 12.1.9).

12.3.3 Feeling on Completion of the Course

Irrespective of the degree programmes, the students felt that about 47 % students are strongly agreed to say I am satisfied with the knowledge, I have gained from the course to start my own farming/ business, 49 % are I am feeling proud to have completed this course, 39 % are I would recommend this course for others. 42 % are agreed that I consider this course at par with other professional degrees. Whereas, 29 % of students are strongly disagree that I regret joining this course, 22 % agree, 23% neither agree nor disagree (Table 12.1.11).

12.3.4 Conduct of Classes and Quality of Teaching

The students opinion regarding conduct of classes and quality of teaching presented here under 47 % are agreed about quality of teaching, 37 % strongly agreed, 36 % are agreed about practical exposure given to students and practical knowledge of teachers, 40% agreed about adequacy of no. of teachers in each subject/course, 36% are agreed about use of AV aids and Multi-media in teaching, 42% are strongly agreed about punctuality in conduct of examinations and coverage of syllabus by teachers (Table 12.1.12).

12.3.5 College Infrastructure and Usage of Library

Maximum students are fully satisfied about the existing infrastructure facilities. On an average 60% students, satisfied with respect to classrooms (space, light, aeration, chairs/benches, AV aids), seminar hall auditorium and library facilities. The usage of library, 46% of students visit library every day, among them 61 % come to read Magazines/Journals/News papers/Periodicals, 53 % come to read text book and prepare notes and 36 % come to read own classroom notes. Students spend on in average 2.14 h per day, 6.89 h per week, and 23.00 h per month (Table 12.1.14).

12.3.6 Skills Acquired /Gained on Completion of the Course

Students expressed their opined about skills upon completion of courses, 34% are very good about time management, 39% committed to work, 40% organized at their work, 33% have good presence of mind (alertness), 40% get well alongwith teachers and co-students, 44 % are excellent in ready to take risk to achieve their goal and 36% good in well-informed about the current affairs (Table 12.1.19).

Table 12.1.1: Joined degree by choice/chance (%)

N = 245

SN	Degree	Choice	Chance	Total
1	UG	68	32	100
2	PG	80	20	100
3	Ph.D.	76	24	100

Table 12.1.2: Reason for joining the degree programme (%)

S.No.	Reasons	UG	PG	Ph.D.
1	Best available option	45	43	33
2	To gain practical knowledge about agriculture to enable me to practice progressive agriculture	44	55	29
3	Job prospects are much better in agriculture career	38	48	38
4	Good course for preparing for Civil Service Examinations	28	33	24
5	Since I did not get any job with my previous qualification	13	5	10

Table 12.1.3: Gender-wise reason for joining the Bachelors degree programme (%)

N = 184

S.No.	Reasons	Male (88)	Female (96)	Total
1	Best available option	56	35	45
2	To gain practical knowledge about agriculture to enable me to practice progressive agriculture	44	44	44
3	Job prospects are much better in agriculture career	39	38	38
4	Good course for preparing for Civil Service Examinations	31	25	28
5	Since I did not get any job with my previous qualification	16	10	13

Table 12.1.4: Gender-wise reason for joining the Masters degree programme (%)

N = 40

S.No.		Male (27)	Female (13)	Total
1	Best available option	52	23	43
2	To gain practical knowledge about agriculture to enable me to practice progressive agriculture	59	46	55
3	Job prospects are much better in agriculture career	56	31	48
4	Good course for preparing for Civil Service Examinations	30	38	33
5	Since I did not get any job with my previous qualification	7	0	5

Table 12.1.5: Gender-wise reason for joining the Ph.D. degree programme (%)

N = 21

S.No.		Male (16)	Female (5)	Total
1	Best available option	38	20	33
2	To gain practical knowledge about agriculture to enable me to practice progressive agriculture	38	0	29
3	Job prospects are much better in agriculture career	44	20	38
4	Good course for preparing for Civil Service Examinations	25	20	24
5	Since I did not get any job with my previous qualification	13	0	10

Table 12.1.6: Opinion about the Course Curriculum considering all degree programmes (%)

N=245

S.No.	Particulars	Opinion/Perception* (%)				
		1	2	3	4	5
	The present course curriculum					
1	Matches requirements of self-entrepreneurship	22	47	21	9	2
2	Matches requirements of job in govt. sector	25	36	26	7	6
3	Matches requirements of job in private sector	23	42	18	13	4
4	Covers more of theory than practical sessions	18	27	32	20	3
5	Is loaded with heavy course work (lot many credits)	22	29	23	17	9
6	Has lot of repetition / overlapping	17	29	27	20	7
7	Needs to include more of field visits	49	23	12	11	6
8	Is more focused on providing information than imparting problem solving skills	27	33	24	7	7

* 1=Strongly Agree, 2=Agree, 3=Neither Agree nor Disagree, 4=Disagree, 5=Strongly Disagree.

Table 12.1.7: Opinion about the Course Curriculum of UG Programme

N=184

S.No.	Particulars	Opinion/Perception* (%)				
		1	2	3	4	5
	The present course curriculum					
1	Matches requirements of self-entrepreneurship	24	43	22	9	2
2	Matches requirements of job in govt. sector	26	36	26	8	5
3	Matches requirements of job in private sector	22	41	20	14	4
4	Covers more of theory than practical sessions	19	26	32	21	3
5	Is loaded with heavy course work (lot many credits)	22	35	18	17	8
6	Has lot of repetition / overlapping	15	29	27	21	8
7	Needs to include more of field visits	47	24	12	10	7
8	Is more focused on providing information than imparting problem solving skills	28	30	26	9	7

* 1=Strongly Agree, 2=Agree, 3=Neither Agree nor Disagree, 4=Disagree, 5=Strongly Disagree.

Table 12.1.8: Opinion about the Course Curriculum of PG

N=40

S.No.	Particulars	Opinion/Perception* (%)				
		1	2	3	4	5
	The present course curriculum					
1	Matches requirements of self-entrepreneurship	18	63	13	5	3
2	Matches requirements of job in govt. sector	30	35	28	8	0
3	Matches requirements of job in private sector	25	48	10	13	5
4	Covers more of theory than practical sessions	18	35	20	25	3
5	Is loaded with heavy course work (lot many credits)	30	8	25	18	20
6	Has lot of repetition / overlapping	28	18	28	20	8
7	Needs to include more of field visits	60	20	3	10	8
8	Is more focused on providing information than imparting problem solving skills	30	45	15	0	10

* 1=Strongly Agree, 2=Agree, 3=Neither Agree nor Disagree, 4=Disagree, 5=Strongly Disagree.

Table 12.1.9: Opinion about the Course Curriculum of Ph.D.

N=21

S.No.	Particulars	Opinion/Perception* (%)				
		1	2	3	4	5
	The present course curriculum					
1	Matches requirements of self-entrepreneurship	10	52	29	10	0
2	Matches requirements of job in govt. sector	10	38	24	0	29
3	Matches requirements of job in private sector	33	38	14	5	10
4	Covers more of theory than practical sessions	14	24	52	5	5
5	Is loaded with heavy course work (lot many credits)	10	14	57	19	0
6	Has lot of repetition / overlapping	10	52	29	10	0
7	Needs to include more of field visits	38	19	29	14	0
8	Is more focused on providing information than imparting problem solving skills	14	43	24	5	14

* 1=Strongly Agree, 2=Agree, 3=Neither Agree nor Disagree, 4=Disagree, 5=Strongly Disagree.

Table 12.1.10: Opinion about the conduct and usefulness of following programmes at UG level

N=245

S.No.	Particulars	Opinion/Perception* (%)					
		Conduct of the programme			Usefulness of the programme		
		1	2	3	1	2	3
1	RAWE/Internship programmes	56	40	4	54	40	6
2	Experiential Learning	48	43	9	50	40	9
3	Field Visits (Farms/Industries/Institutions)	43	41	16	51	37	12
4	State tour	62	27	11	49	34	17
5	All India Tour	65	25	10	50	34	16
6	Inter-Collegiate Youth Festival a) Cultural	41	44	16	47	31	21
	b)Literary	46	38	16	47	35	18
7	Inter-University Youth Festival a) Cultural	41	44	16	42	38	20
	b)Literary	42	43	14	49	34	17
8	Inter-Collegiate Sports Meet	46	36	18	45	35	20
9	Inter-University Sports Meet	42	40	17	43	36	22

1=Fully Satisfied, 2=Somewhat Satisfied, 3=Not Satisfied

Table 12.1.11: Perception about/feeling on completion of the course

N=245

S.No.	Particulars	Perception* (%)				
		1	2	3	4	5
1	I am satisfied with the knowledge I have gained from the course to start my own farming/business	47	35	10	6	2
2	I am feeling proud to have completed this course	49	37	9	3	2
3	There are good prospects of getting a job after graduation	33	37	19	8	4
4	General public considers this course at par with other professional degrees	32	37	20	9	3
5	I consider this course at par with other professional degrees	28	42	21	5	4
6	I regret joining this course	15	22	23	10	29
7	I would recommend this course for others	39	33	20	5	4

* 1=Strongly Agree, 2=Agree, 3= Neither Agree nor Disagree, 4=Disagree, 5=Strongly Disagree

Table 12.1.12: Opinion regarding conduct of classes and quality of teaching

N=245

S.No.	Particulars	Opinion* (%)				
		1	2	3	4	5
	Quality of teaching					
1	Knowledge upgradation of faculty in respective field	37	32	19	10	2
2	Practical exposure given to students	30	36	20	10	3
3	Practical knowledge of teachers	34	36	20	4	7
4	Adequacy of no. of teachers in each subject/course	22	40	21	9	8
5	Use of AV aids and Multi -media in teaching	25	36	22	10	7
6	Interaction between students and teacher in the class	27	31	25	12	5
7	Approachability and friendliness of teachers	25	37	22	10	7
8	Punctuality of teachers in taking class	33	33	22	9	4
9	Evaluation/examination system	29	33	20	8	11
10	Complete coverage of syllabus	25	36	24	8	7
11	Punctuality in conduct of examinations	42	33	12	7	4

* 1=Fully Satisfied, 2=Somewhat Satisfied, 3=Neither Satisfied nor Dissatisfied, 4=Somewhat Satisfied, 5=Fully Dissatisfied.

Table 12.1.13: Opinion about College Administration

N=245

S.No.	Particulars	Opinion* (%)				
		1	2	3	4	5
1	Functioning of Placement Cell/Campus recruitments	23	28	13	7	29
2	Fund availability for students' co-curricular activities	9	37	19	17	18
3	Fund availability for students' academic activities	13	33	24	14	15
4	College administration (Supportive/Encouraging)	21	34	22	15	8
5	Functioning of Alumni Association	22	29	27	13	9
6	Functioning of Students' Association	24	27	24	15	9
7	Present students' admission procedure	24	40	18	13	4
8	Present fee structure	16	36	22	11	14
9	Safety & security of students in the campus	33	34	17	10	6
10	Treatment of students (gender, racial, regional discrimination)	29	29	20	12	9

* 1=Fully Satisfied, 2=Somewhat Satisfied, 3=Neither Satisfied nor Dissatisfied, 4=Somewhat Satisfied, 5=Fully Dissatisfied.

Table 12.1.14: Opinion of the students about college infrastructure

N=245

S.No.	Particulars	Opinion* (%)				
		1	2	3	4	5
a)	**Curricular:**					
1	Classrooms (space, light, aeration, chairs/benches, AV aids)	43	35	10	7	4
2	Laboratories (Space, number & quality of equipments)	37	32	18	9	5
3	Conference/Seminar Hall (Space & equipped)	41	31	14	10	4
4	Accessibility to Computers and accessories	32	33	13	16	7
5	Accessibility to Internet/Email	29	29	18	14	11
6	Accessibility to photocopying facility	22	36	18	14	10
7	Vehicles for field visits (Adequacy and quality)	28	25	18	17	12
8	Library (availability of books, journals, news papers, *etc.*)	62	22	8	4	4
b)	**Supporting amenities:**	0	0	0	0	0
1	Hostel accommodation	32	33	14	13	8
2	Mess/Canteen/Food facilities	24	33	20	13	9
3	Outdoor sports facility	29	31	19	12	9
4	Indoor sports facility	21	29	23	15	13
5	Cultural room+ instruments	18	25	22	17	18
6	Gymnasium hall	15	31	24	13	17
7	Auditorium	36	34	13	10	8
8	Medical service (Clinic)	25	33	16	14	13
9	Medical Shop	13	25	17	16	29
10	Bank	40	24	11	8	16
11	Post Office	39	31	11	4	14
12	Grocery shop	22	28	15	9	27
13	Swimming pool	13	16	8	12	52

* 1=Fully Satisfied, 2=Somewhat Satisfied, 3=Neither Satisfied nor Dissatisfied, 4=Somewhat Satisfied, 5=Fully Dissatisfied.

Table 12.1.15: University Library usage by the students

S.No.	Frequency of visits	No. of students	% of Students
1	Every day	112	46
2	Weekly	71	29
3	Fortnightly	20	8
4	Monthly	12	5
5	Occasionally	30	12
6	Purpose of visit		
	a) To read own notes for exams	87	36
	b) To refer textbooks for exams	129	53
	c) To access internet	90	37
	d) To read Magazines/Journals/News-papers/Periodicals	149	61
	e) To search information for assignment	96	39
	f) To meet friends	18	7
	g) Others (specify)	1	0
7	Average time spend in library		0
	hr/Day	2.14	1
	hr/Week	6.89	3
	hr/Month	23.00	9
8	Average do you allocate for co-curricular activities (hrs per week)		0
	hr/Day	1.84	1
	hr/Week	5.56	2
	hr/Month	12.44	5

Table 12.1.16: Skills acquired / gained on completion of the course (all UG and above degree students)

N=245

S.No.	I am / I*	Opinion* (%)				
		1	2	3	4	5
1	Good at time management	31	34	22	10	2
2	Innovative	29	31	27	10	3
3	Sincere	32	33	25	9	2
4	Hard working	30	31	26	10	3
5	Committed to work	25	39	25	10	2
6	Punctual	29	35	25	8	3
7	Organized at my work	28	40	22	9	1
8	Confident in writing skills	27	34	27	11	2
9	Confident in oral presentation	29	30	29	10	2
10	Well informed about the subject	22	31	33	11	3
11	Well informed about the current affairs	18	25	36	18	4
12	Improved oral presentation skills	25	30	30	13	2
13	Improved computer skills	21	29	32	15	3
14	Improved written presentation skills	23	31	31	12	2
15	Improved leadership skills	29	28	28	13	2
16	Have stage courage	27	27	30	11	4
17	Have good presence of mind (alertness)	28	33	28	9	2
18	Have good decision making skill	27	36	26	10	1
19	Get along well with teachers	22	40	25	11	3
20	Get along well with co-students	29	41	22	7	1
21	Ready to take risk to achieve my goal	44	30	17	8	1
22	Usually follow rules of the University/College	34	33	23	8	1

* 1=Excellent, 2=Very Good, 3= Good, 4=Moderate, 5= Poor

Table 12.1.17: Will you start your own business after some years

S.No.	Response	Bachelors (No.)	Masters (No.)	Doctoral (No.)
1	Yes	23	7	1
2	No	161	33	20
		184	40	21

Table 12.1.18: What are the factors/constraints hindering your career preference (all UG and above degree students)

N=245

S.No.	Particulars	Rank										
		1	2	3	4	5	6	7	8	9	10	11
1	Medium of study at higher secondary level	49	42	22	23	11	14	12	13	7	3	12
2	Medium of study at XII standard level	33	35	33	21	16	12	13	7	11	17	3
3	Guidance/support from family, relatives & friends	41	41	18	31	21	14	12	5	14	12	7
4	Absence of career planning by me	75	33	34	12	8	11	10	15	7	3	7
5	Low OGPA	30	41	23	18	20	17	14	9	7	12	11
6	Family type and Family background	34	30	25	40	22	4	18	6	8	17	10
7	Education of parents	18	31	26	23	10	24	12	25	13	17	0
8	Income of parents- high	33	20	26	17	10	11	19	16	21	13	6
9	Occupation of parents	25	37	17	20	25	5	9	20	18	15	10
10	Mass media exposure in parents	11	24	27	20	15	18	16	13	18	30	7
11	Socio-political participation of parents	9	15	23	24	17	17	6	11	14	11	40

Table 12.1.19: Will you continue higher studies in the same University

S.No.	Response	Bachelors (No.)	Masters (No.)	Doctoral (No.)
1	Yes	95	16	0
2	No	89	20	3
3	Not Applicable	0	4	18
		184	40	21

Table 12.1.20: What are your expectations from this University

S.No.	Expectations	Bachelors (No.)	Masters (No.)	Doctoral (No.)
1	Campus Placement	127	26	13
2	Help to mobilize Credit/Finance	55	12	7
3	Help in start-up/business incubation	55	9	9
4	Coaching for competitive exams	84	14	11

12.4 Survey of Teachers in Agriculture and Allied Sectors

The teaching profession is one of the most visible professions in the world, and even though significant improvements have been made in student achievement, society continues to expect more from its teachers. As the gap widens between the public's expectations of education and the teacher's ability to deliver that education, quality will continue to be a concern.

The results of teacher's survey showed that, average workload of teachers and scientists in SAUs. Irrespective of cadre in teaching UG 4.55 h, PG 3.44 h and research as a PI handling 1 project & Co-PI 2 Projects. Whereas, an average as a leader or Co-leader involvement in extension is about 2 (Table 12.2.1). The actual time allocation of teachers is 45, 25, 20 and 10 % Teaching, Research, Extension and Corporate/Admin, respectively. But teachers are somewhat satisfied about the actual time allocation. Hence, as per teachers and scientists opinion the desirable is 50, 25, 20 and 5 % Teaching, Research, Extension and Corporate/Admin, respectively (Table 12.2.2 and 12.2.3).

Satisfaction level with respect to the HRD opportunities: 52 % of the faculties are fully satisfied about the permission & support to participate in seminars/conferences/symposia within University or India (55%) and abroad (22%). 64 % of faculties are fully satisfied about the permission to attend the workshop, 56 % to the training. Whereas, about 44 % of faculties are neither satisfied nor dissatisfied about consultancies (Table 12.2.4).

Satisfaction level with respect to physical facilities & staff at office: 58% of faculties are fully satisfied about office workspace and ambience, classroom (50 %), presentation facilities (54%). still faculties are responded that, they are fully dissatisfied in term of documentation (20 %), Vehicle and POL (18%), Technical staff adequacy (18%) and supporting staff adequacy (20%) Table 12.2.5.

Amenities available/accessible in the campus/college: Teachers are rated the amenities available/accessible to them in the campus/college, such as lift facilities, special sitting arrangements in the classroom, special AV-Aids, Staff quarters, medical service (clinic), medical shop, post office, canteen/cafeteria, outdoor sports facility, indoor sports facility (72, 67, 63, 52, 72, 45, 54, 67, 66, 73 %) are rated as good. Still them some are expressed the poor facilities in terms of vehicle facilities, special treatment during field visit and lab work, swimming pool, *etc.* (Table 12.2.6).

Opinion about students, examinations, evaluation and grade announcement: the students efficacy in terms of punctuality/response/ interaction of students during class is the 36 % are excellent, 46 % faculties responded that class strength of students (intake) is very good, and also quality of students, quality of presentation in UG courses, students' relationship with teachers and students' behaviour outside class/campus (52, 44, 50, 60 and 48 %) is very good, respectively (Table 12.2.7).

Students' quality attributes: The quality of the students are rated by the teachers *viz.*, students are very good in sincerity (44%), committed to work (46%), get along well with teachers (42%), have good presence of mind (52%), have good leadership quality (48%), usually follow rules of the University (40%). Whereas, students are good in hard working (40%), confident in oral presentation (50%), get along well with co-students (40%), get along well with teachers (44%) and have good decision-making skill (50%). About 30 and 34 % are moderate in have good leadership quality and usually follow rules of the University (Table 12.2.8).

Perception about University/College Administration/Programmes: Teachers perception about administration with respect to sufficient funds available; for teaching (42%), research (56%), extension (68%), students' activities (48%) are strongly agree. The administration is supportive/encouraging in university and college (56 and 54 %) are agreed, respectively and also agreed about students' association is functioning effectively (50%), good job prospects for our graduates/ post-graduates (62%), present curriculum matches job requirements (56%) and present fee structure is appropriate (52%) Table 12.2.9.

Efficacy of various activities/events: The efficacy of RAWE programme fully satisfied (48 and 30%); experience learning is somewhat satisfied (66 and 50 %) about conduct and usefulness of the programmes. Similar responses were observed with respect to tour and extra-curricular activities (Table 12.2.10).

Selection criteria for recruiting the graduates: The 70 per cent of the teachers are score first rank for academic qualifications, followed by 14 per cent to inter-personal and communication skills and passion/ knowledge of industry / drive / commitment/ attitude (Table 12.2.11).

Table 12.2.1: Average workload of Teachers and Scientists in SAUs

N= 50

Cadre	Teaching (h/week)		Research Project (no.)		Extension Activities (no.)	
	UG	PG	PI	Co-PI	Leader	Co-leader
Assist. Prof	5.00	2.00	0.53	2.00	1.00	2.00
Assoc. Prof	4.91	3.70	1.60	1.79	1.64	2.06
Professor	3.75	4.62	2.60	1.81	2.64	1.06
Average	**4.55**	**3.44**	**1.57**	**1.87**	**1.76**	**1.71**

Table 12.2.2: Average actual time allocation for Teaching / Research / Extension / Admin. Activities

S.No.	Activities	Actual (%)	*Satisfaction level (%)				
			1	2	3	4	5
1	Teaching	45	16	30	3	1	0
2	Research	25	10	26	13	1	0
3	Extension	20	5	35	8	1	1
4	Corporate/Admin	10	7	25	9	5	4

* 1=Fully Satisfied, 2=Somewhat Satisfied, 3=Neither Satisfied nor Dissatisfied, 4=Somewhat Dissatisfied, 5=Fully Dissatisfied

Table 12.2.3: Average desirable time allocation for Teaching / Research / Extension / Admin Activities by the faculty

N= 50

S.No.	Activities	Desirable (%)
1	Teaching	50
2	Research	25
3	Extension	20
4	Corporate/Admin	5

Table 12.2.4: Satisfaction level w.r.t. the HRD opportunities available to faculty

N= 50

S. No.	Particulars	*Level of Satisfaction (%)				
		1	2	3	4	5
1	**Seminars / Conferences / Symposia**					
	a. Permission & Support for organization of the event	44	44	10	0	2
	b. Permission & Support to participate within University	52	30	14	4	0
	c. Permission & Support to participate within India	52	28	10	8	2
	d. Permission & Support to participate in events abroad	22	54	16	6	2
2	**Workshops**					
	a. Permission & Support for organization of the event	64	22	12	2	0
	b. Permission & Support to participate within University	52	40	4	4	0
	c. Permission & Support to participate within India	56	20	14	10	0
	d. Permission & Support to participate in events abroad	30	34	22	6	8
3	**Trainings**					
	a. Permission & Support for organization of the event	56	34	8	2	0
	b. Permission & Support to participate within University	46	34	12	8	0
	c. Permission & Support to participate within India	48	32	16	4	0
	d. Permission & Support to participate in events abroad	16	48	24	10	2
4	**Consultancies**					
	a. Received by you from other experts	24	28	44	4	0
	b. Provided by you to clients	18	42	26	10	4
	c. Resource generation from providing consultancies	12	32	42	12	2

* 1=Fully Satisfied, 2=Somewhat Satisfied, 3=Neither Satisfied nor Dissatisfied, 4=Somewhat Dissatisfied, 5=Fully Dissatisfied

Table 12.2.5: Satisfaction level w.r.t. physical facilities & staff at office

N= 50

S. No.	Particulars	*Level of Satisfaction (%)				
		1	2	3	4	5
1	Office workspace and ambience	58	26	14	2	0
2	Office furniture (adequacy and quality)	48	42	6	2	2
3	Class rooms (Sufficiency and quality)	50	28	20	2	0
4	Laboratories (Space and Equipments)	32	36	22	10	0
5	Conference Hall	36	46	10	4	4
6	Computer & accessories in your office/accessible	30	50	12	4	4
7	Photocopying facility (easy & timely accessibility)	40	38	16	4	2
8	Internet (easy & timely accessibility)	32	34	26	6	2
9	Presentation facilities (LCD Projector & Screen)	54	42	0	0	4
11	Light & Sound System	32	36	32	0	0
12	Documentation facility (Camera, Video, etc.)	34	24	16	6	20
13	Vehicle/POL for field visits	26	42	8	6	18
14	Technical Staff (adequacy & quality)	38	18	18	8	18
15	Supporting Staff (adequacy & quality)	32	18	20	10	20
16	Computer stationery (toner, CD, paper, etc.)	46	12	30	8	4
17	Office stationery (writing papers, gum, pencil, etc.)	32	34	12	2	20

* 1=Fully Satisfied, 2=Somewhat Satisfied, 3=Neither Satisfied nor Dissatisfied, 4=Somewhat Dissatisfied, 5=Fully Dissatisfied

Table 12.2.6: Rate the amenities available/accessible to you on the campus/ college

N= 50

S. No.	Particulars	Percentage of faculties says Yes	Opinion*			Percentage of faculties says No
			1	2	3	
1	Lift facility	36	17	72	11	64
2	Special sitting arrangements in the classroom	54	26	67	7	46
3	Special AV -Aids	76	16	63	21	24
4	Staff quarters	96	19	52	29	4
5	Medical service (Clinic)	94	15	72	13	6
6	Medical Shop	62	19	45	35	38
7	Bank	86	47	40	14	14
8	Post Office	96	35	54	8	4
9	Canteen / Cafeteria	96	17	67	17	4
10	Grocery shop	74	27	24	49	26
11	Outdoor sports facility	94	30	66	4	6
12	Indoor sports facility	96	19	73	8	4
13	Cultural room+ instruments	86	28	42	30	14
14	Gymnasium hall	92	37	37	26	8
15	Auditorium	94	62	28	11	6
16	Swimming pool	66	42	24	33	34
17	Vehicle facility	90	42	22	36	10
18	Special treatment during field visits/ lab work, etc.	80	40	30	30	20

*1=Excellent, 2=Good, 3 = Poor

Table 12.2.7: Opinion regarding students, examinations, evaluation and grade announcement

N= 50

S.No.	Efficacy of various Activities	*Opinion (%)				
		1	2	3	4	5
1	Punctuality/response/interaction of students during class	36	16	16	32	0
2	Class strength of students (intake)	20	46	16	2	16
3	Students' attendance in the class	30	34	34	2	0
4	Quality of students (merit)	20	52	18	8	2
5	Students' interest in extension activities	12	24	46	18	0
6	Submission of practical records/ assignments	12	38	40	8	2
7	Quality of presentation in UG courses	10	44	36	8	2
8	Quality of PG seminars given	42	28	14	12	2
9	Students' relationship with teachers (Cordial)	16	50	28	4	0
10	Students' behaviour outside class/campus	18	60	18	4	0
11	Students' behaviour in the exam hall	20	34	42	4	0
12	Students' behavior in co-curricular activities	12	48	36	4	0

* 1=Excellent, 2=Very Good, 3=Good, 4 = Moderate 5 = Poor

Table 12.2.8: Students' quality attributes as perceived by the teachers

N= 50

S.No.	Students/are	*Quality (%)				
		1	2	3	4	5
1	Sincere	10	44	36	4	6
2	Hard working	16	28	40	14	2
3	Punctual	26	28	20	24	2
4	Committed to work	10	46	18	24	2
5	Innovative	20	26	44	6	4
6	Organized in their work	8	36	22	30	4
7	Good at time management	20	18	24	34	4
8	Confident in oral presentation	20	14	50	8	8
9	Confident in writing skills	8	36	22	32	2
10	Well informed about the subject	26	22	40	12	0
11	Well informed about the current affairs	20	20	32	18	10
12	Have stage courage	20	24	40	12	4
13	Get along well with co-students	10	32	44	14	0
14	Get along well with teachers	12	42	40	6	0
15	Have good presence of mind (alert)	4	52	34	6	4
16	Have good decision-making skill	8	24	50	18	0
17	Have good leadership quality	18	48	22	12	0
18	Usually follow rules of the University	12	40	36	10	2
19	Ready to take risk to achieve their goal	14	28	46	6	6
20	Any other (specify):					

* 1=Excellent, 2=Very Good, 3=Good, 4 = Moderate 5 = Poor

Table 12.2.9: Perception of teachers about University/College Administration/ Programmes

S. No.	Students/are	*Opinion (%)				
		1	2	3	4	5
1	Sufficient funds are available for teaching activities	34	42	4	20	0
2	Sufficient funds are available for research activities	22	56	18	2	2
3	Sufficient funds are available for extension activities	8	68	18	6	0
4	Sufficient funds are available for students' activities	26	48	14	8	2
5	University administration is supportive / encouraging	30	56	8	4	2
6	College administration is supportive / encouraging	22	54	20	4	0
7	Colleagues in the dept/section are supportive	54	20	8	12	6
8	Supporting staff in the dept/section are supportive	38	38	18	4	2
9	Placement Cell is functioning effectively	22	38	30	6	4
8	Safety & security in college and campus is good	18	46	10	8	18
9	Teachers' Association is functioning effectively	26	24	12	16	22
10	Students' Association is functioning effectively	6	50	18	8	18
11	Alumni Association is functioning effectively	14	34	20	10	22
12	Good job prospects for our graduates / post-graduates	16	62	16	4	2
13	Present curriculum matches job requirements	26	56	12	6	0
14	Present curriculum loaded with more theory than practical's	16	20	48	14	2
15	Curriculum not focused on imparting problem solving skills	18	22	36	20	4
16	Present fee structure is appropriate	20	52	20	2	6
17	Present students' admission procedure is appropriate	26	36	32	0	4
18	Staff recruitment procedure is appropriate and periodical	14	28	12	12	30

* 1=Strongly Agree, 2=Agree, 3=Neither Agree nor Disagree, 4=Disagree, 5= Strongly Disagree

Table12.2.10: Opinion of teachers about efficacy of various activities/ events in the University

S. No.	Particulars	Opinion/Perception* (%) Conduct of the programme			Opinion/Perception* (%) Usefulness of the programme		
		1	2	3	1	2	3
1	RAWE programme	48	30	22	30	48	22
2	Experiential Learning	30	66	4	46	50	4
3	Field Visits (Farms/Industries/ Institutions)	36	40	24	38	40	22
4	State tour	46	30	24	40	36	24
5	All India Tour	52	28	20	44	38	18
6	Inter-Collegiate Youth Festival a) Cultural	32	56	12	42	56	2
	b) Literary	30	60	10	34	60	6
7	Inter-University Youth Festival a) Cultural	26	70	4	48	50	2
	b) Literary	44	54	2	50	46	4
8	Inter-Collegiate Sports Meet	44	52	4	50	44	6
9	Inter-University Sports Meet	52	46	2	44	50	6

* 1=Fully Satisfied, 2=Somewhat Satisfied, 3=Not Satisfied

Table 12.2.11: Employees opinion about selection criteria for recruiting the graduates

S. No.	Employability Skills	Rank									
		1	2	3	4	5	6	7	8	9	10
1	Interpersonal and Communication Skills (%)	14	18	18	22	4	18	4	2	0	0
2	Academic Qualifications (%)	70	20	0	6	0	2	0	2	0	0
3	Work Experience (%)	4	32	28	20	10	2	4	0	0	0
4	Leadership Skills (%)	2	22	8	10	34	22	0	0	0	2
5	Passion/ Knowledge of Industry / Drive / Commitment/ Attitude (%)	14	40	2	4	18	10	2	6	2	0
6	Teamwork Skills (%)	6	6	12	16	6	8	38	6	0	2
7	Critical Reasoning and Analytical Skills/ Problem Solving/ Technical Skills (%)	8	14	34	6	4	12	8	10	0	2
8	Emotional Intelligence - including self-awareness, strength of character, confidence, motivation (%)	4	4	4	18	2	0	20	36	6	4
9	Activities - includes both intra- and extra-curricular (%)	6	6	2	0	12	2	0	14	50	6
10	Cultural Alignment/ Values Fit (%)	4	6	0	0	0	12	0	0	18	58

12.5 Survey of Working Graduates in Agriculture and Allied Sectors

Agriculture professionals working in public and private sector for them interpersonal skills are incredibly important. They are routinely required to interact with officials, dealers, and farmers, *etc.* Hence, agriculture graduates should do problem-solving and able to adjust all situations. They have to take more and more personality development and business courses to understand the environment in which they will practice their theory. If longer in wait to teach such aspects, the greater are the chances of losing our critical mass in terms of relevancy and support in terms of soft skills along with knowledge. Hence, graduates can have broad range of knowledge towards his career and clientele groups. Thereby graduates can immediately adjust with the corporate world which leads to increase in efficiency.

The results of working graduates survey showed that, the present activity of the working graduates is 74 % are employed on salary, 13 % are self-employed and 13 % looking for work (Table 12.3.1). In total

9% from Govt. Developmental Departments, 23% financial institutions, 14% from fertilizer industries, 14% pesticides industries, 23% seeds sectors and 18 % from NGOs. The average salary of working graduates in Govt. Developmental Departments Rs. 47,000, financial institutions Rs. 35,000, fertilizer industries Rs. 43,000, pesticides industries Rs. 40,000, seeds sectors Rs. 38,000 and NGOs Rs. 25,000. Whereas, degree wise average experience and salary is 4 year and Rs. 18,000 for diploma holders, 8 years and Rs. 35,000 for UG, 6 year and Rs. 45,000 for PG and 4 years and Rs. 54,000 for Ph.D. Details were presented in Table 12.3.2 and Table 12.3.3.

Working graduates were asked that how they are selected for present employment; about 50 % are direct selected in interview through personal contact, 22.73 % are through open job advertisement, 18 % are through campus interview and 9.09 % are through other competitive exams (Table 12.3.4). After completing the degree nearly 73 % of graduates took less than 6 months to get job, 18 % are got job immediately after competing degree, whereas 9 % graduates took more than six months (Table 12.3.5). Whatever the knowledge get through agriculture education is about only partially adequate as opinioned by 73 per cent of graduates (Table 12.3.6).

The self-employing graduates were asked that reason for choice of self-employment is because of own interest (100%). Various problems faced are raw materials (50%), other inputs (25%) and marketing (25%).

About 90 per cent of the working graduates selected agriculture/ allied sector degree as their own interest, 7 per cent are due to more employment opportunities and 3 per cent are due to no specific interest in subject, but a professional degree is useful for competitive examinations. About the course curriculum, 100 % of the graduates say yes to more interaction with industries. Whereas, more advanced theoretical knowledge, more technical knowledge and more practical work (83, 77 and 93 %) said yes, respectively.

Table 12.3.1: Present Activity of Graduates

N=30

S.No.	Activity	Gender		
		M (%)	F (%)	Total
1	Employed on salary	73	27	74
2	Self-employed	100	0	13
3	Looking for Work	75	25	13
	Total	**100**		

Table 12.3.2: Survey of working graduates in different organizations

N=22

S. No.	Organisation	Percentage of Persons	Average year of Experience	Average salary (Rs.) per month
1	Govt. Devp. Dept	9	9	47000
2	Financial	23	7	35000
2	Fertilizer	14	5	43000
4	Pesticides	14	4	40000
5	Seeds	23	5	38000
6	NGOs	18	4	25000
	Above all	**100**	**6**	**38000**

Table 12.3.3: Prevailing average salary per month

N=22

S.No.	Degree	Average Experience(years)	Average Salary (Rs.) per month
1	Diploma	4	18000
2	UG	8	35000
2	PG	6	45000
4	Ph.D.	4	54000
	All	**5.5**	**38000**

Table 12.3.4: Chance of getting employment by agricultural graduates

N=22

S.No.	Time Taken	Responses
1	Campus interviews	18.18
2	Other competitive examinations	9.09
3	Open job advertisements	22.73
4	Direct interviews*	50.00

*through open web advertisement, application through email or website, reference by teachers, seniors and friends

Table 12.3.5: Average time required to get employment after completing studies

N=22

S.No.	Time Taken	Responses of Working graduates (%)
1	Immediately	18
2	Less than six months	73
3	6 months to 1 year	9
4	1 year or more but less than 2 years	0
5	2 years or	0

Table 12.3.6: Adequacy of education from university to get the job

N=22

S.No.	Adequacy	Responses (%)
1	Fully adequate	17
2	Partly adequate	73
3	Not at all	10

Table 12.3.7: Reasons for choosing self-employment (%)

N=4

S.No.	Adequacy	Responses
1	Own interest	100
2	Family business	0
3	Lack of salaried jobs	0
4	Better opportunities for growth	0
5	If other	0

Table.12.3.8 : Different problems faced by self-employed graduates

N=4

S.No.	Problems	Responses (%)
1	No problems	0
2	Raw materials	50
3	Power	0
4	Other inputs	25
5	Marketing	25
6	Technical knowhow	0
7	Lack of demand for the product/service	0
8	If others	0

Table 12.3.9: Reason for joining agricultural/allied sciences degree (as opinioned by working graduates)

N=30

S.No.	Reason	Responses
1	Interest in the subject	90
2	Best degree as per my qualification	0
3	More employment opportunities in sector	7
4	Family pressure	0
5	No specific interest in subject, but a professional degree is useful for competitive examinations	3
6	If others	0

Table 12.3.10: Working graduates suggestion for improving course curriculum

N=30

S.No.	Problems	Percentage of graduate say	
		Yes	No
1	More advanced theoretical knowledge	83	17
2	More technical knowledge	77	23
3	More practical work	93	7
4	More interaction with industries	100	0
5	Managerial skills	73	27
6	Information technology	63	37
7	Agri-business economics	57	43
8	If any other, please specify	0	0

12.6 Institution/Organization/Establishment Survey

The organization or institution is an economic unit that produces and/or sells goods or services, and operates from a single physical location. If a firm has several such locations, each is termed as an establishment. In order to make an establishment profitable the skilled human recourse is mandatory. In agriculture and allied sector now-a-days lot of changes has been taken place still the sector has to grow further to reach expected level of growth.

The 30 establishments were covered in the survey indicates that they employed on an average, 50 agricultural graduate per establishment. Among the establishments engaged in different activities related to agriculture and allied sciences that, in developmental department had the largest number employed (500 per establishment the state as a whole). Financial establishments have much higher average (200 per establishment) as the banks provided data at state level.

Stakeholders were asked to indicate the growth in their industries/ organizations and employment over the previous five years, *i.e.,* from 2010-11 to 2014-15. Data shows the pattern of change in employment. But few of the private organization did not provided the data for all the five years employment growth. If that was the case, the adjustment was made by taking the sectoral growth rate. It is seen that the total employment has increased at a compound annual growth rate of 10 per cent per annum over the five years period.

The data which respect to education qualification indicates that, about 60 were with general qualifications and remaining 40 were with some qualifications in agriculture and related fields. Among the agriculture graduates, 57 per cent under-graduates, 31 per cent post-graduates, 2 per cent doctorates and the rest 10 per cent being below under-graduate level.

The results showed that some of the private establishments are engaged, the graduates other than the agricultural graduates in the place of agricultural graduates work. This is mainly due to the agriculture graduates demanding more salary than non-agriculture graduates, compared to agriculture graduates non-agriculture graduates are more stick on the job and more importantly agriculture graduates demanding more salary and not able to do ground level work. Finally, they opinioned that graduate with non-agriculture profession are more significant than agriculture graduates.

Although it is mainly due to the fact that every establishment engages non-professional personnel for various activities such as administrative, financial, *etc.,* it is also probably due to shortage of agriculturally qualified personnel, their high turn-over and high remuneration demanded by them. It may also be noted that a substantial percentage (10 per cent) was with informal training and diploma holders, indicating the demand for agriculture professionals in future. Highly qualified (post-graduates and doctorates) are preferred for research and they are less likely to be hired by establishments for field or general work.

Data on distribution of employees engaged in various types of major activities by their disciplines, indicates that, major field of activity of the establishment, the agriculture graduates features prominently in all activities in all the sectors mainly in private sector.

It is observed in the establishment survey that, private sector were not agreed to disclose their expansion plans and future requirements, *etc.* Establishments indicated their human resources requirements for 2024-25. In the public sector, due to governments policies, the establishments did not foresee much growth even if there is a need for human resources. Data clearly showed that, 30 out of 50 establishments provided this information. It is seen that the total employment of graduates and above in agricultural and allied sciences, as envisaged by the 5 % growth at present employment, is likely to increase from about 10 % in 2024-25 due to expansion, diversification and retirement. The information related to the vacancies arising due to retirements, transfers, expansion, *etc.* and the extent to which they were filled. Generally, there is delay in filling vacancies in public sector. It is seen that about 10 to 15 per cent of the total number of positions in establishments appear to fall vacant each year and about 50 per cent vacancies remain unfilled in the past five years.

Majority of the establishments (95 %) indicated that they would expand the organizations, diversify products, introduce technological changes and modernize the organizations. The organizations showed lot of interest towards the skilled human resources requirements in future for the technological change, innovations and organizational modernization of existing system finally product diversification with existing investment to win the competitive market.

The 75 per cent of the private organizations expressed that there is skill gap in the agriculture graduates. Somewhat technically, they are excellent but many of the well-qualified post-graduates also lacking soft skills (Table 12.4.1). Then establishments have to give special training to boost up the soft skills. They were of the opinion that to meet the corporate sectors requirements, students should have more advanced theoretical as well as technical knowledge with practical orientation and has to know how to express the available technical knowledge. This requires more interaction between educational institutions and industries during graduation level. The graduates have to require more and more managerial and information technological skills to work efficiently (Table 12.4.2).

Table 12.4.1: Employers opinion about agriculture and allied sector graduates to meet the skill requirements

S.No.	Adequacy	Responses (N=30)	Responses (%)
1	Fully	9	30
2	Partly	21	70
3	Not at all	0	0

Table 12.4.2: Future skill requirements as opined by employers (%)

N=30

S.No.	Skills	UG level	PG level
1	More practical orientation	67	77
2	More technical knowledge	83	67
3	More advanced theoretical knowledge	67	60
4	Managerial skills	50	60
5	More interaction with organizations (During studies)	90	73
6	Agri- business economics	70	33
7	Information technology	40	57
8	Communication skills	17	10

12.7 Disciple-wise Quality, Skill issues and Institutional Mechanisms

12.7.1 Agriculture, Agriculture marketing and Agri. Engineering, Home Science, Food Technology

The students survey revealed that 68 percentage of UG, 80 percentage of PG and 76 percentages of Ph.D. students joined degree by choice. Many of them are expressed that they joined degree by choice because this degree is the best available option for them. Most of the UG students (>70%) preferred to join higher studies (post-graduation) and about 30 % are showed interest to join Ph.D.

There is a great demand for practical-oriented agricultural education from industries as well as from students. About (75%) of graduates working in the industries are frankly expressed that what they learned/ gained knowledge from their education is partly adequate to hold the job.

Teachers expressed that presently substantial component of practicals was already there in the syllabus, but students and industries felt that the emphasis on practical's was not enough to make students employable. Hence, at graduation level itself student's needs to involve more in practical oriented aspects as a course curriculum. It is necessary to provide the students more practical orientation from time to time so that they could have mastery over some subjects as well as development of soft skills. There is a need to limit theory classes and introduce other methods such as group discussions, case studies, presentations by students, quizzes, interaction with experts, conferences, and research paper presentations and so on.

Experts and industries opined that students are lack of in-depth knowledge of subjects. It is better to attach students with farms and industries for a few days, the training at mid of the graduation instead of final year or fag end of their education. Hence, it helps students to discuss with faculty after knowing the problems of industries.

Major reason for the lack of skills and knowledge of students is the depleted faculty strength, lack of manpower in frontier areas, inadequate hands-on skills and lack of research experience, which leads to lot of work load for the existing faculty.

It is observed that non-government sector; private corporate sector in future will emerge as major employer for agriculture graduates. They expressed that the students, who take up employment in agriculture sector, leave the job as soon as they get a better employment even if it is not in their specialized field. This problem could be addressed treating agriculture at par with other professional courses to make the sector attractive to professionals. Hence, Human resources retention is of concern to some employers. Private companies and organizations employing many basic or other diploma/degree holders, even for positions requiring Agriculture and allied sector graduates, for continuity in business as many of the agriculture graduates frequently leave the job and demanding high salary as compared to other general degree graduates. They prefer to have more diploma holders for routine works in their various field related work and other operations. Industries would like to employ Agriculture graduates in preference to general graduates for many of their requirement, if the agriculture graduates are ready to do all type of work.

12.7.2 Horticulture

From the students survey, it was observed that some of the lacunae in horticulture education such as large number of courses were taught at the under-graduate level enabling the students to pick up only a little information about everything. No subject was taught in-depth which could give specific skills to the students in at least a few areas. Curriculum was not practical oriented (means not much industry oriented). Presently, students are attached with industries, farms and other organizations only for a few days. The training at these places remains limited to observation only. Hence, they hardly got less practice in the farms or industries. Better to provide the students more practical orientation from time to time so that they could have mastery over some subjects as well as skills. It was also pointed out that the syllabus of graduation should be revised involving industries, students and faculty together.

The number of courses should be reduced with specialization in certain areas. Courses also needed to include applied statistics, advanced biochemistry, communication and managerial skills, personality development, mathematics, as well as recent development aspects of the subjects.

Recent develop in various aspects of production, processing and marketing are including in the practical and they get translated into practice. This requires a multi-faceted approach. For example, the new technologies are to be publicized on a continuous basis, the students should get an opportunity to practice the new technology and its implications, the mindset of the farmers has to be changed so that they could apply the new technologies on their farms. Programmes like farm-industry linkages and university-industry-farm linkages are must.

Most critical problem facing is the lack of trained and expert teachers. Even where courses have been modified and made practice oriented, teachers have not been fully trained. In a University, a substantial number of positions are vacant and there is dearth of teachers.

It has been observed that more number of urban background students getting admitted in this disciple. They take the course as merely a degree and not a professional course to excel further. Therefore, after obtaining a degree, they try for MBA courses and other courses other than horticulture. A suggestion was made that the students who own land and belong to farmer community, should be given a preference at the time of admission so that they could opt for horticulture as their career. Another alternative could be that agriculture is introduced as a subject at school level like other subjects and students with real aptitude for the sector would develop interest and opt for this sector later not by compulsion but by choice.

The horticulture sector should be treated as an industry and specific concessions should be given to agro-based industries as was done for IT sector in order to encourage the self-employment or entrepreneurship.

12.7.3 Forestry

Forestry education requires practical orientation and constant interactions with industries/organizations. Students have to understand the basics subjects' clearly *viz.*, plant biology, taxonomy, physiology, *etc.* and also have a knowledge of new areas such as aroma courses, remote sensing and GIS, *etc.*

The emerging areas for employment in this field are - forest certification, environmental science, wood and paper technology, products from medicines and herbs, nursery management, plantation management, forest protection, climate impact, carbon credits, *etc.*

Molecular biology techniques are yet to be used in forestry propagation. A beginning has taken place and this may emerge as a strong field in the future. Clone forestry is growing fast, particularly in plantation companies.

There is no focused forestry activity with respect to pasture, grass land management in forest area. Hence, there is a lot of demand for trained skilled human resource in this area to work at ground level. As stakeholder opined that limited job opportunities in the private sector to the forestry graduates makes forestry degree program is not much attractive for students. Majority of graduates opting for higher studies, as they are not getting suitable jobs after graduation. Therefore, a scope for employment opportunities in agro processing (wood and paper) industries products by providing advanced training and knowledge in these industry is thereby making students to compete with national or international level.

Forest Development Authority and Forest Monitoring Agencies need additional human resources at least one in each centre, considering the recent public interest on mining and plantation activities. There is need for stronger monitoring mechanism from the public system. This needs additional human resources apart from state forest department.

For processing industries, environmental issues are very important and consider they are most critical, if polluted the air, water and other resources. Therefore, the industries like paper mills need post-graduates in Paper Technology as well as in Environmental Science. The graduates are absorbed by Industry to deal with the issues of environment, wood and paper technology development and to limited extent on plantation development. It has been suggested by the students that forestry should be introduced as a separate subject in Civil Services Examinations.

12.7.4 Sericulture

The survey and focus group discussions during the study showed that a high percentage of students preferred to join higher studies after graduation *i.e.* most of them tried to go for post-graduation. It has been reported that quality of students joining to the sericulture education is coming down. Students are more attracted towards agriculture courses as compared to sericulture courses. Thus, students who join sericulture, generally do not come by choice.

In order to attract talented students to the course, there is need to revised curriculum with addition of more practicals on advanced sericulture technology. Basic foundation for Self-Entrepreneurship development has to begin at college. Also courses like project planning, finance, *etc.* to be taught more effectively. Need to focus on creation and make available the quality teaching material and need to promote books & teaching aids for rhe benefit of teachers and students.

As like in other agriculture and allied sectors in sericulture also great demand for practical-oriented education from industries as well as from students. Many sericulture graduates working in the public and private are expressed that what they learned/gained knowledge from their education is partly adequate to hold the job. Therefore, the course curriculum should include industry orientation-internship, emphasis on innovative research, and have strong industry interaction, Public Private Partnership are necessary to improve the quality of the education. In addition to meet the grassroot level requirements, skilled personnel, need to offer short-term specialised training of 2-3 months to be useful at micro level on the lines of finishing school concept.

12.7.5 Animal Husbandry and Veterinary Science

Qualified and trained people are needed, who can handle the practical aspects relating to livestock such as artificial insemination, embryo transplant, and vaccination of animals, etc., with a professional approach. At present, the students coming out of the veterinary education, have inadequate practical orientation.

There is a need for emphasis on development of quality human resources in the following sectors to meet the future growth and challenges in the sector. The poultry industry has tremendous potential for growth but there is a dearth adequate human capital to make the poultry industry profitable, similarly lack of scientific slaughter houses for the pork industry in the state. The state has considerable goat and sheep population but milk is not being used properly. New areas have emerged in this sector such as impact of climate change on animal health and production of milk production. In our country or state does not have a proper breeding policy or fodder development policy. In spite of the higher rates of milk production, milk yield is very low as compared to other world average. If any improvement in this sector can make big changes in terms of white revolution is concern.

The demand for veterinary graduates would increase, as there would be more self- employment and also additional requirement for more veterinary clinics. At present, the Veterinary & Animal Husbandry education is not producing entrepreneurs. Students lack business abilities and communication skills. Veterinary & Animal Husbandry course should be able to develop entrepreneurial skills. Therefore, there is need for revision veterinary education curriculum with inclusion of management IT skills, leadership quality, *etc.*, so that the students with veterinary education could have better options for enhancing employment in both core, veterinary sector and related organisations.

There is a shortage of faculty in all veterinary colleges due to non-availability of qualified candidates since students after passing out from the universities prefer to join Corporate Sector, go abroad, try for competitive exams, join MBA, *etc*. One of the reasons for this phenomenon is low salaries and prestige to the sector. There is also a need for attracting talented and retraining of existing faculty with special incentives.

In order to provide the veterinary service in rural area, it needs a more number of diploma and certificate holders.

There is need for focus on specialized skilled manpower in the following areas: the development of milk products by milk processing, besides increasing shelf life of milk and value addition, adequate supply chain, production losses, low extent of processing and fortification of milk, milk production, management of intellectual property rights, patenting and trade, technological innovation and research, production as per international standards, equipment design and maintenance, and advanced food packaging.

There is a demand of skilled graduates for food security, food quality, food safety and total quality management. The entry of corporate sector in research, the demand for post-graduates in dairy technology has increased to the extent of 10 per cent.

12.7.6 Fisheries

It is evident from student's survey that most of the students joined to the fisheries course are by chance. They joined fishery because they didn't got veterinary so. Even after completion of under-graduate in fisheries, they go for higher studies like MBA or they try for jobs in Middle East and other African countries, where there is a demand for fishery educated personnel with higher incomes.

There is a grater scope for self-employment in the state but presently only less than one per cent of the professionals in fisheries are getting attracted towards the self-employment. There is a need to motivate the students passing out from fisheries colleges to take up self-employment in seed production, growth promotion, fish feed, processing and storage industries in the state. In addition, there is a scope to establish super markets in the rural areas, which are potential employment of fisheries. There is requirement of introduction of entrepreneur and business development and communication skills courses in final year of degree program.

There is need for introduction and upgradation of the marine engineering courses, since marine aquaculture and navigation are the major areas of fishery sector, in which lot of demand for skilled human resources are requirement to improve marine fishing technology.

The fishing activities are majorly taken by uneducated person/ farmers in rural areas, the researches going on in the sector should be disseminated in a simple language so that farmers can understand the new innovations. Hence, along with the routine course curriculum, it is very much essential to include communication skills, training on local languages, *etc.*

In the state, fishery department is having shortage of staff due to that various schemes are not reaching to the farming communities. Similarly, shortage of teaching faculty in Fishery College, which leads to quality of education is hampered. Therefore, there is need for govt. policy for regular and timely recruitment in Universities and Department.

There is suggestion from a student's that, in order to attract the more number of talented students towards fisheries sector. Presently, a fishery is not a separate subject in the civil services examination. There is a demand that this subject should be included as a separate subject for the competitive examinations.

The diploma holders in fisheries are lot of opportunities in the different enterprises like boat management and fish processing, *etc.*, are employing general graduates. Hence, there is a demand for diploma holders because they can work at cheaper rates.

12.8 The expected advanced knowledge and skill set needed

Several concerns have been expressed for declining quality of agricultural graduates. Major concerns are depleted faculty strength, lack of manpower in frontier areas, inadequate hands-on skills and lack of research experience. The students frankly expressed their views about the improvements, they would like make in the curricula and teaching.

12.8.1 Agricultural Sector (including Agri. Marketing, Agri. Engineering, Agri. Biotech and Food tech, *etc.*)

- Soil health and nutrient management (Integrated Nutrient Management, Site specific nutrient management, Precision farming, *etc.*)
- Advanced tools and techniques for Pest and Disease Management.
- Standard procedure/protocols for quality control during production, processing, storage and marketing of agricultural produce and Seed production, grading, testing and quality control.

- Hands on training on Agri-Business entrepreneurship with latest Knowledge of e–Marketing.
- Integrated Farming system and Organic Farming and its importance.
- Recent developments in micro irrigation, liquid fertilizers, fertigation and greenhouse technologies for conservation of natural resources.
- Skills for farm machineries operation, repair and maintenance hi-tech equipments.
- Development equipments and its testing as per standard protocol and also maintenance of machinery and equipments.
- Food processing and value addition to the Agriculture produce
- Recent advances and knowledge on Natural Resource management-watershed development and importance of groundwater recharge, Development and utilization of wasteland and use of renewable energy resources for agriculture.

12.8.2 Horticulture Sector

- Knowledge of recent technology *viz.*, Biotech and tissue culture, greenhouse technology, grafting, drip irrigation, packing technology, fruit processing and cold storage, *etc.*
- Techniques of preservation of fruits and vegetables and value addition.
- Basics and applied aspects of nursery management, raising of tray nursery, grading of produce, produce quality testing, *etc.*
- Diversification in farming or integrated farming and high density planting.
- Hi-tech horticulture and precision farming.
- Advanced technologies in Pre-& Post-harvest management of fruits, vegetable and flower quality.
- Entrepreneurship development in horticulture business.
- Recent advances in floriculture, olericulture and pomology.
- Commercialization and business knowledge about medicinal and aromatic plants and their processing and packaging.

12.8.3 Veterinary and Animal Husbandry Sector

- Super specialists for surgery in different livestock and wild
- Management of animal health and animal nutrition, and also hygiene and safety
- Milk processing and value addition to milk products and processed foods marketing skills
- Biotechnological approaches for Animal breeding, artificial insemination and also embryo development
- Specialists for Goat and sheep rearing, pest & Disease control
- Fodder production and storage for longer period.

12.8.4 Fisheries Sector

- Integrated development of fish and fish products
- Storage of fish products for longer period
- Value addition to fish by processing and packaging
- Fish disease management
- Knowledge of ornamental fish, shrimps, hatcheries, prawn and their development
- Production of nutritive fish feed.

12.8.5 Forestry Sector

- Forest produces processing (wood and paper technology)
- GIS and remote sensing application in forest management
- Detail knowledge of forest taxonomic classification
- Establishment of forest in wasteland, landscape gardening and decrees deforestation
- Creation of awareness to farmers to plant trees around field and near to home
- Maintenance of forest nursery and supply of good quality plating material
- Collection, identification, processing and preservation of endanger forest seed species.
- Effective management of forest service centres.

12.8.6 Sericulture Sector

- Advanced mulberry cultivation and improvement of productivity
- Different species of worms and new techniques in rearing of worms
- Skill and knowledge for sericulture business management including international silk and silk products trade.
- Advanced method of Extraction and quality testing of silk
- Hygiene and safety measures in rearing room
- Seri-Biotechnology.

Chapter **13**

Conclusions and Recommendations

The sector specific issues have been discussed in respective chapters in detail. This chapter would focus upon certain key results, conclusions and policy options.

The present study "assessment and estimation of human resource requirements in agriculture and allied sectors in Karnataka". The earlier chapters provided the results of forecasts made on the basis of growth pattern in each sector and sub-sectors in the light of various developmental programmes and the vision for each sector. A primary survey undertaken as a part of the study has provided the basic inputs needed to finalize the human resource needs with their required skill sets according to various educational levels. The study looked at the human resource requirements not only in quantitative terms but also on issues relating to quality as well. It also considered the institutional mechanisms for imparting education in agriculture and its allied sectors and addressed skill related issues in the context of changing labour market scenario as required in present day agriculture. Reviewed the various scheme and programmes available for HRD of agricultural and allied sectors students in the state.

13.1 Major Findings

- There is an increased trend in supply of agriculture and allied sectors graduates in Karnataka. The annual compound growth rate of supply of agriculture and allied sectors graduates in Karnataka during 2010-2011 to 2014-15 was 10.19 per cent. Considering the same trend in supply (CGAR 10.19 %), the estimated supply of agriculture and allied sectors graduates for 2024-25 is about 8,813.
- It has been observed that, there is a diversified employment opportunities for agricultural allied sector graduates in private sector as compared to public sector. The estimated demand for

agricultural and allied sectors graduates for 2024-25 is 13,277. Hence, there is a 34 per cent gap in demand for and supply of agricultural and allied sectors graduates for 2024-25 in Karnataka.

- The student's survey of agricultural and allied sectors indicates that most of the students expressed that there is a need for revised curriculum of course towards more effective practical oriented than theory.
- The stakeholder survey (employers) of agriculture and allied sectors reveals that there is a need for improving in technical, managerial and communication skills of graduates. In addition, there is a strong opinion from all stakeholders for skill -specific education with clarity of basics as well as hands on technical expertise.

13.2. Policy Options

- The results show that there are at present substantial gaps between demand and supply of human resources in agricultural and allied sciences even to the tune of 34 per cent or more. This is true across the sectors, though the shortfall is high in the case of rapidly growing veterinary, fisheries, forestry and horticulture sectors and less serious in others. Dairy, fisheries and horticulture are the future engines of growth in the agricultural sector and are important from the point of view of food security of the state and nation. Their growth, in particular, should not be hampered by human resources unavailability snags.
- The study also shows that the present educational facilities at their current levels of output will not be able to cope with the demand pressures likely to emerge in the next decade and substantial step up is required even to contain or slightly reduce the current and future demand-supply gap. No doubt, there has been some acceleration in the out turn of high level agricultural human resources in recent years, but it has been shown that even this belated accelerated growth would hardly make a dent on the seriousness of the problem. The situation demands much stronger initiatives. Perhaps an increase of 2 to 3 times in disciplines like veterinary, fisheries, forestry and horticulture.
- Apart from numbers, planning for agricultural education in the next decade has to look at the structure of the system. Till

recently, the emphasis in the system has been more or less exclusively on the generation of graduates, post-graduates and doctorates and very little effort has been put in generation of intermediary skills at diploma level. Initiatives for the future demand-driven growth in agricultural education at the diploma and under-graduate levels needs to be given due importance.

- An important aspect of educational planning is to ensure availability of faculty in adequate numbers and quality with institutional arrangements for systematic up-gradation of their knowledge and skills. It has been observed that about 40 per cent of the faculty positions in the farm universities have remained vacant for long periods of time raising questions about the quality of education. On the other hand, it has also been observed that at post-graduate level, the teacher-pupil ratio is extremely low in many disciplines, calling for a better organization of teaching system. It has been observed that at post-graduate level, the intake per course is very low as compared to other fields of technical and professional education. There is a need to consolidate the post-graduate courses (like crop science, crop protection Social Science, etc.) to conserve the resources for optimum utilisation.

- One of the critical factors in agriculture in the Karnataka dominated by small and marginal farmers is one of inadequate transfer of technology and consequently low productivity. Therefore, there is a need for creating/strengthening extension system by single window system by both online and offline for providing day-to-day advisory services and assistance to the farmers at their doorstep. In addition to providing/rendering routine farm practices, veterinary services, testing services, and various other rural occupations but also on aspects like agro-processing, marketing, *etc.*

- An important development, in agricultural education in recent years, has been the increasing participation of girls, with their share of enrolment reaching about 40 per cent overall. The development of soft skills and other entrepreneurship skill courses can be imparted to them so that they could be economically active.

- There is need for creating and maintenance of centralized database of all education, research and extension activities at University level. This database would help for identification and prioritization of research, education and extension activities at university level, state and national level. Data Analysis Cell (Data Science) needs to be established conduct research on farmers and other stakeholders problems in addition to support for education planning.
- Quality of education is directly related to the employability. Consultations with industry and academicians as well as students suggested a large gap between the skills imparted in the institutions and skills required in the employment market. This calls for a re-look into the curriculum, teaching strategies and establishing forward and backward linkages. Although, the Fifth Deans Committee, mandated to revise agricultural curriculum, recommended a strong practical orientation to agricultural education. The issues relating to course syllabus, effective teaching strategies, and quality of teachers need serious thinking and policy interventions while considering expansion of agriculture education as well as meeting the labour market challenges.
- There is a strong demand from all the stakeholders for skill-specific education with clarity of basics as well as hands-on technical expertise. In other words, there is a need to develop functional skills among the students in educational institutions almost in all the sectors of agriculture.
- Although the results of the study show that human resources supply in almost all the sectors of agriculture is less than the demand and therefore, unemployment is low but as reported the agri-education is not considered as a professional course at par with other professional courses, especially at graduate level. Pro-active policies are needed to give agriculture its proper dignity and place. This would also facilitate in attracting youth to join the sector not as a compulsion but as a choice.
- Agriculture is undergoing rapid transformation with the advent of new technologies, commercialization and globalization and emergence of issues relating to climate and environment. Teachers have to keep pace with these developments and the resulting need to acquire new knowledge. While some

organisations are imparting training in the emerging areas, there is no specific training policy for up-gradation of the skills of teachers. Therefore, a plan of action has to be evolved so as to cover all the teachers under capacity development programme in a specific period of time. Need to create the teacher training units in the state similar that of HAU, Haryana and NAARM.

- Adequate institutional funding, and infrastructure and effective management are some of the areas that directly made an impact on quality of education. Focus should be made towards upgradation and increase the infrastructure facilities existing and new colleges. Available budgetary resources are far from satisfactory to develop infrastructure or world class laboratories in the colleges and universities.

- Regular appointment in development departments is strongly recommended so that extension services in the state should be strengthened on priority with adequate infrastructure facilities, staff and other resources to cater to the needs at micro level. Regular follow-ups of the programmes operated under extension services are also needed. Synergies in various organisations such as Extension services, NGOs, SHGs, farmers associations, *etc.* are extremely essential.

- There is need for introducing the agriculture subject at high school level to motivate students towards agriculture and allied sectors, as many students are shying away from this sector, Madhya Pradesh state already has been introduced the agriculture courses at the school level.

❐❐❐

References

Alexandratos N., Bruinsma J., 2012. World agriculture towards 2030/ 2050: the 2012 revision. ESA Working paper No. 12-03. Rome, FAO.

Anonymous, 2005, Agricultural Human Resource Development. ICAR, Annual Report, 2005. pp. 1-5.

Anonymous, 2005, Status of agriculture education in India (from the data collected by ICAR to prepare the report on agricultural education for meeting of the Consultative Committee of the Ministry of Agriculture), Education Division, ICAR, New Delhi.

Anonymous, 2013, Agricultural Human Resource Development. DARE/ ICAR, Annual Report, 2012–13. pp. 73-74.

Anonymous, 2006. Current Trends in Occupational Employment and forecasts for 2010 and 2020, Final report to The Expert Group on Future Skill Needs, The Economic and Social Research Institute.

Anonymous, 2017, Economic survey of Karnataka, 2016-17, Published by Department of Planning, Programme Monitoring and Statistics, Govt. of Karnataka.

Anupam Varma, 2014, Agricultural Education in India: Imaging Possibilities to Meet Challenges in the Changing World. *National Agricultural Education Day Lecture* IARI, New Delhi, November 13, 2014. p.10.

Ashish.K. Makwana, 2013, Agricultural education in India: challenges and prospects. *Voice of Research*, 2 (3): 2277-7733.

Ayyappa, S., Pitam Chandran and Tandon, S. K., 2007, ICAR-Industry Meet Agricultural Transformation through Public-Private Partnership: An Interface. ICAR report, pp. 1-13.

Balagopal G Menon1, Biswajit Mohanty, D and Rama Rao, 2010, Assessment of Technical Manpower Requirements in Agriculture Sector in India, Annual Report published by NAARM, Hyderabad.

Bisht, B.S., Vizayakumar, K., and Sahu, K.C., 1996, Soyabean Oil Processing Industry in India - Policy Perspectives, System Dynamics: *An Int. J. of Policy modelling*.

Bullock, K.W, and Litzenberg, K.K., 2002: Forecasting Future employment Opportunities for Food, Agriculture, and Natural Resources Higher Education Graduates Using Adjusted Bureau of Labour Forecasts, Department of Agricultural Economics, Texas A&M University

Chung, S.H., Jung, D.C., Yoon, S.N. and Lee, D.H., 2010, "A dynamic forecasting model for nursing manpower requirements in the medical service industry", *Service Business*, Vol. 4, pp. 225-236.

CIFE, 2000, Manpower Requirements and Human Resource Development in Fisheries Sector, Education Division of Indian Council of Agricultural Research and Central Institute of Fisheries Education, Mumbai.

FAO, 1970, Agricultural Manpower Planning, Training and Utilization, Country Profile Study - India. FAO, Rome.

Forrester J.W., 1961, Industrial Dynamics, MIT Press, Cambridge, MA.

Forrester, 1961, Agricultural Education in India in "Higher Education in India". Ed. Powar, K.B. and Panda S.K., Association of Indian Universities, New Delhi.

Franklin, J.C.,2007, Employment Outlook: 2006-16, An Overview of BLS Projections to 2016, Monthly labour review – November 2007, United States of America, Department of Labour, Bureau of Labour Statistics:

Hinchliffe, K., 1993: Human Resources\Forecasting and Rapid Labour Market Analysis: The demand for technical school graduates in Egypt, Journal of Vocational Education & Training, Volume 45, Issue 3, 1993

Jantsch, E. (1972), "Forecasting and the systems approach: A critical survey", *Policy Sciences*, 3: 475-498.

Khoong, C.M. 1996, "An integrated system framework and analysis methodology for manpower planning", *International Journal of Manpower*, 17(1): 26-46.

Mandal, P., 1993. A system dynamic model for food production. *An Int. J. of Policy Modelling* 6 (1): 5-25.

Mandal, P., Mohapatra, P.K.J., Jain, A., and Sangh, V.K.C., 1992.Analysis of scientific manpower flow scenarios for an R&D organization. *J. of Sci. and Industrial Research*, 51 : 685-692.

Martino, J.P., 1980, "Technological forecasting-an overview", *Management Science*, 26(1):28-33.

Mohapatra,P.K.J., Mandal,P., and Bora, M.C., 1994. Introduction to System Dynamics Modelling, University Press, Hyderabad.

Nanda, S. K., Rama Rao, D. and Vizayakumar, K., 2004, Human resource development for agricultural sector in India: a dynamic analysis. NAARM report, pp. 1-12.

Parker, B. and Caine, D., 1996, "Holonic modelling: human resource planning and the two face of janus", *International Journal of Manpower*, Vol. 17, No. 8, pp. 30-45.

Parker, B. and Caine, D., 1996, Holonic modelling: human resource planning and the two face of janus", *International Journal of Manpower*, 17(8): 30-45.

Parnes, H.S., 1962: Forecasting Educational Needs for Economic and Social Development, OECD, Paris.

Poultry Sector, Central Avian Research Institute, Izatnagar.

Rama Rao, D., Nanda, S. K. and Ram Prasad, 2011, Assessment of Future Human Capital Requirements in Agriculture and Allied Sectors, NAARM Annual Report, 2010-11. p. 198.

Ramarao, D., R. V. Kumari and E. Haribabu, 2000, Agricultural education in India: a sociological perspective. *Outlook*, 29: 177-184.

Rashmi Agrawal, Awasthi, I. C., Joshi, G. P., Sanchita Bhattacharya and Indra Kumar, 2011, Assessment of Future Human Capital Requirements in Agriculture and Allied Sectors, IAMR Annual Report, 2010-11. pp. 208-209.

Roy, D. and Koul, S., 2009, "Human resource planning in a shore-based integrated steel plant: A system dynamics model", *Proceedings of the 27th International Conference of the System Dynamics Society*, Albuquerque, USA.

Sandra Damico, 2010, Higher Education and Skills for the Labour Market in Cambodia. Sustaining Rapid Growth in a Challenging Environment, Cambodia Country. *Economic Memorandum*, World Bank, January 2009, p.71-74.

Sasidhar, P.V.K., 2009, Poultry Science Education and Human Resource Planning.

Shivanagaraju, C., Mahanty, B., Vizayakumar, K., and Mohapatra, P.K.J., 1998. Beta- distributed age in manpower planning models, *Applied mathematical Modelling*, 22: 23-37.

Stepp *et al.*, 2009, Department of Animal Husbandry and Dairying, Annual Report, 2007-08

Stepp, M.D., Winebrake, J.J., Hawker, J.S. and Skerlos, S.J., 2009, "Greenhouse gas mitigation policies and the transportation sector:

The role of feedback effects on policy effectiveness", *Energy Policy*, 37: 2774-2787.

Sterman, J.D., 2000, Business dynamics: systems thinking and modeling for a complex world. Irwin, Chicago

Suresh Sinha, 2000, Education for agriculture in India: Time for a change. *Current Sci.*, 79:302-310.

Swaminathan, M. S., 2005, Redefining agricultural education and extension system in changed scenario. NAAS, New Delhi, Executive Council Meeting report. p. 3.

Tamboli, P. M. and Nene, Y. L., 2013, Modernizing Higher Agricultural Education System in India to Meet the Challenges of 21st Century. *Asian Agri-History*, 17 (3): 251–264.

Vinay *et al.*, 2016, Fisheries education in Karnataka: Trends, current status, performance and prospectus. *Journal of World Academy of Science, Engineering and Technology. 10(10): 3306-3310.*

Vinay G., 2010, Indian seed Industry – An Overview, www.commodity India.com, 16 Nov., 2015

Vizayakumar.K., 1990. A study on some aspects of environmental impact analysis. Ph.D. Thesis, Department of Industrial Engineering and Management, Indian Institute of Technology, Kharagpur.

Wolstenholme and Coyle, 1983, Manpower Forecasting and Modelling Replacement Demand: An Overview. Research Centre for Education and the Labour Market. ROA-W-1996/4E, http://edocs.ub.unimaas.nl/loader/file.asp?id=528 .

Wu, M., Xu, Q., Gui, B., Zhu, L. and Chen, J., 2003, "System dynamics modelling of manpower forecasts and programming", *Proceedings of the 21st International Conference of the System Dynamics Society*, New York, USA.

*Annual Reports of the all farm universities and developmental departments Govt. of Karnataka, from 2010-11 to 2014-15.

□□□

Annexures

Annexure-1: List of Agricultural Universities and Colleges in Karnataka

S.No.	University College
I.	**University of Agricultural Sciences, Bangaluru (May 25, 1963)**
1	College of Agriculture, Bangalore
2	College of Agriculture, Mandya
3	College of Sericulture, Chintamani
4	College of Agriculture, Hassan
II.	**University of Agricultural Sciences, Dharwad (October 1, 1986)**
1	College of Agriculture, Bijapur
2	College of Agriculture, Dharwad
3	College of Agriculture, Hanumanamatti
4	College of Forestry, Sirsi
5	College of Rural Home Science, Dharwad
III.	**University of Agricultural Sciences, Raichur (April 1, 2009)**
1	College of Agriculture, Raichur
2	College of Agriculture , Bheemarayanagudi
3	College of Agricultural Engineering, Raichur
4	College of Agriculture, Kalaburagi
IV.	**Veterinary, Animal and Fisheries Sciences University, Bidar (February 23, 2004)**
1	Veterinary College, Bangalore
2	Veterinary College, Bidar
3	Veterinary College, Shimoga
4	Veterinary College, Hassan

5	Dairy Science College, Bangalore
6	Dairy Science College, Gulbarga
7	College of Fisheries, Mangalore
8	Animal Husbandry Polytechnic, Shiggaon
9	Animal Husbandry Polytechnic, Tiptur
V.	**University of Horticultural Sciences, Bagalkot (Nov., 22, 2008)**
1	College of Horticulture, Bagalkot
2	College of Horticulture, Bidar
3	College of Horticulture, Arabhavi
4	College of Horticulture, Munirabad
5	College of Horticulture, Sirsi
6	College of Horticulture, Kolar
7	College of Horticulture, Mysore
8	College of Horticulture, Bangalore
VI.	**University of Agricultural and Horticultural Sciences, Shimoga (Sept. 21, 2012)**
1	College of Agriculture, Shimoga
2	College of Horticulture, Mudigere
3	College of Forestry, Ponnampet
4	College of Horticulture, Hiriyur
5	Agricultural Diploma Institute, Kathalagere

Annexure-2 : Employment Opportunities for Agriculture and Allied Sector Graduates in Public and Private Sectors

S.No.	Sectors	Opportunities
1	Agriculture	Agriculture Dept. (40%) and other public Dept.- Co-operative Department, RDPR, KSRTC, Administrative posts (10%) and R&D and Education (5%), Financial institutions (20 %), Seed industry (5%), Fertilizer industry (5%), Farm equipment (2%) and Pesticide industry (5%), Self-employment (0.5%), NGOs (5%) and others (2.5%).
2	Agril. Engineering	Agriculture Dept. (4%) and other public Depts.- RDPR, KSRTC, Administrative posts (5%) and R&D and Education (5%), Tractor companies (40%), Financial institutions (5 %), Farm equipment and processing industry (20%), Self-employment (2%), NGOs (5%) and others (5%).
3	Agril. Marketing	Agriculture Marketing & Co-op dept. (20%) and other public dept.-RDPR, Administrative posts (10%) and SAU's (5%), Financial institutions (20%), agriculture input sector (30%), Self-employment (0.5%), NGOs (4%) and others (10.5%).
4	Agril. Bio-Technology	Tissue culture lab (30%), Veterinary Pharmaceutical Sector (20%), Academic Sector (5%), NGOs (5%), Teaching (3%), Research (5%), Extension (1%), Public institutions (2%), Seed companies (8%), and others (21%).
5	Rural Home Science	Developmental Departments (40%), SAUs (8%), Food Processing Industries (15%), Hospitals (2%), Self-Employment (2%), NGOs (5%),Financial Institutions & Insurance companies (7%) and Others (29%).
6	Food Technology	Developmental Departments (2%), SAUs (5%), Food Processing Industries (35%), Self-Employment (2%), NGOs (15%), Financial Institutions (5%) and others (36%).
7	Horticulture	Horticulture Departments (40%), SAU's-Education, Research and Extension (10%), Seeds/ saplings-nurseries/seed farms (20%), Agro

		Chemical Industries (10%), Financial institutions (3%), Post-harvest- quality control, transport, cold storage (10%), NGOs (2%) and others (5).
8	Veterinary	Veterinary Dept. (40%), R&D (7%) and Education (5%), Public Veterinary Services (5%), Private Veterinary Services (15%), Financial institutions (2%), Vet. Pharmaceutical and animal drugs (10%), Animal feed industries (2%), Milk and Meat Processing Industries (5%), Self-Employment (2%), NGOs (2%) and others (5%).
9	Dairy Technology	Veterinary Departments (15%), SAU's (10%), KMF& Co-operative societies(10%), milk and milk processing industries (25%), Self-Employment (2%), NGOs (10%) and others (28%).
10	Fishery	State Fisheries Department (45%), SAU's (5%), Financial Institutions (5%), Processing plants (5%), Fish/Shrimp seed hatcheries (5%), Shrimp/Fish rearing units (5%), Feed mills (5%), Canning industries (2%), aquaculture farms (4%), aqua-feed plants (5%), deep sea fishing vessels (1%), and pharmaceutical companies (2%), NGOs (7%), Self Employment (2%) and others (2%).
11	Forestry	Forestry Department (60%), SAU's (8%), Administrative posts (5%), Financial Institutions (1%), Agro Processing- Timber & Wood Industry (10%), NGOs (2%), Self-Employment (0.5%) and Other (13.5%).
12	Sericulture	Sericulture Department (50%), Silk Board (5%), SAU's (8%), Sericulture Processing Industry (10%), Financial Institutions (4%), NGO's (10%), Self Employment (3%) and others (10%).

Source: Establishment survey, discussion with experts, teachers and institutional heads, *etc.*

Annexure-3

Annx. 3.1: Land-use and Irrigation in Karnataka

S. No.	Classification	Year				
		2010-11	2011-12	2012-13	2013-14	2014-15
1	Total Geographical Area			190.50		
2	Net Area Sown	105.23	99.41	97.93	99.23	101.34
3	Area sown more than once	25.40	21.18	19.55	23.44	21.13
4	**Total Cropped Area**	**130.62**	**120.59**	**117.48**	**122.67**	**122.47**
4A	**Area under Agricultural Crops**	111.62	101.23	98.74	103.94	102.26
5	***Cropping Intensity - %***	**124**	**121**	**120**	**124**	**121**
6	Net Irrigated Area	34.90	34.40	34.21	35.56	36.59
7	Net Irrigated area in %	33	35	35	36	36
8	Gross Irrigated Area	42.79	41.37	40.07	41.12	41.86
9	Gross Irrigated Area in %	33	34	34	34	34
10	**Cropping Intensity in Irrigated areas**	**123**	**120**	**117**	**116**	114
11	Net rain fed area	70.33	65.01	63.73	63.68	64.75
12	Net rain fed area in %	66.83	65.39	65.07	64.17	63.90
13	Gross rain fed area	87.83	79.23	77.41	81.55	80.60
14	Gross rain fed area in %	67.24	65.70	65.89	66.48	65.82
15	**Cropping Intensity in Non-Irrigated areas (rain-fed areas)**	**125**	**122**	**121**	**128**	**124**

Source: ASCR of DE&S.

Annx. 3.2 Per cent changes in area of Agriculture and Horticulture crops during last 10 years

		2005-06						
S.No.	Crop group	Area	Percent	Area	Percent	Area	Percent	Growth Rate (%)
Agriculture								
1	Cereals	56.16	50	48.73	48	-7.43	-13	-1.15
2	Pulses	20.50	18	23.21	23	2.71	13	1.48
3	Oilseeds	28.63	25	13.72	14	-14.91	-52	-8.17
4	Commercial crops	8.25	7	15.39	15	7.16	87	6.45
	Total	**113.54**	**100**	**101.04**	**100**	**-12.49**	**-11**	**-1.15**
Horticulture								
1	Fruits	2.58	14	4.07	18	1.49	58	4.96
2.	Vegetables	4.03	22	4.63	21	0.59	15	0.91
3.	Spices	2.32	12	2.09	9	-0.23	-10	-2.61
4	Flowers	9.56	51	11.22	50	1.66	17	1.66
5.	Other*	0.23	1	0.34	2	0.12	50	4.35
	Total	**18.72**	**100**	**22.35**	**100**	**3.64**	**19**	**1.60**
	Grand total	**129.36**		**123.39**		**-5.97**	**-5**	**-0.64**

Source: Economic survey, Govt of Karnataka, 2015-16.

Annexure-4: Employment Growth in Public and Private sectors

Annex. 4.1.1: Employment growth rate in private agriculture sector

Organizations	2011	2012	2013	2014	2015	CAGR (%)
Financial Institutions	52	17	31	39	45	5.6
Seed sector	10	17	13	26	32	31.7
Fertilizer sector	24	15	16	33	46	23.2
Pesticide sector	14	17	22	28	35	26.3
Farm Machineries	12	16	27	28	33	29.5
NGO's	4	12	9	9	15	26.6
Others	12	17	15	18	22	13.5
Total	**128**	**111**	**133**	**181**	**228**	**17.9**

Source: Compiled from placement cells of all SAU's

Annex. 4.1.2: Employment growth rate in private Horticulture sector

Organizations	2011	2012	2013	2014	2015	CAGR (%)
Financial Institutions	2	1	3	1	3	8.4
Seeds Industry	4	3	5	6	6	16.2
Fertilizers &Pesticides Industry	2	1	3	3	2	11.6
Processing and Marketing sector	3	2	4	2	7	18.5
Total	**11**	**7**	**15**	**12**	**18**	**16.5**

Source: Compiled from placement cells of horticulture universities and colleges

Annex. 4.1.3: Employment growth rate in private Veterinary sector

Organizations	2011	2012	2013	2014	2015	CAGR (%)
Vet. Pharmaceuticals	3	3	9	9	7	32.2
Poultry Industry	3	5	5	7	9	28.8
NGO's	2	2	3	3	6	29.7
Others	3	2	4	4	5	18.7
Total	11	12	21	23	27	27.7

Source: Compiled from placement cell of veterinary university

Annex. 4.1.4: Employment growth rate in private Fishery sector

Organizations	2011	2012	2013	2014	2015	CAGR (%)
Hatcheries	2	2	5	2	8	32.0
Fish Processing	2	4	5	2	7	19.9
Feed Industry	1	1	1	1	3	24.6
NGO's	1	3	1	5	2	20.9
Others	1	1	1	3	2	28.2
Total	7	11	13	13	22	27.9

Source: Compiled from placement cell of Fishery College

Annex. 4.1.5: Staff position in public sectors (farm universities) as 31.3.2015

UAS, GKVK, Bangaluru					
S.No.		**Sanctioned**	**Filled**	**Vacant**	Vacancies %
1	Professor	56	33	23	41.07
2	Associate professor	153	106	47	30.72
3	Assistant professor	401	285	116	28.93
	Total	**610**	**424**	**186**	**33.57**
UAS, Dharwad					
1	Professor	57	18	39	68.42
2	Associate professor	90	43	47	52.22
3	Assistant professor	448	329	119	26.56
	Total	**595**	**390**	**205**	49.07
UAS, Raichur					
1	Professor	40	2	38	95.00
2	Associate professor	65	5	60	92.31
3	Assistant professor	363	255	108	29.75
	Total	**468**	**262**	**206**	**72.35**
UAHS, Shimoga					
1	Professor	46	21	25	54.35
2	Associate professor	67	24	43	64.18
3	Assistant professor	236	127	109	46.19
	Total	**349**	**172**	**177**	**54.90**
UHS, Bagalkot					
1	Professor	80	61	22	23.75
2	Associate professor	125	95	30	24.00
3	Assistant professor	214	160	51	25.23
	Total	**419**	**316**	**103**	**24.33**

KVAFSU, Bidar					
1	Professor	87	38	49	56.32
2	Associate professor	99	44	55	55.56
3	Assistant professor	209	128	81	38.76
	Total	**395**	**210**	**185**	**50.21**
Grand total					
1	Professor	366	173	196	52.73
2	Associate professor	599	317	282	47.08
3	Assistant professor	1871	1284	584	31.37
	Total	**2836**	**1774**	**1062**	**43.73**

Annex. 4.1.6: Staff position in public sectors (developmental departments) as 31.3.2015

Department of Agriculture					
S.No	**Group**	**Sanctioned**	**Filled up**	**Vacant**	**Vacancies %**
1	**Group –A**	609	466	143	23.48
2	**Group –B**	4006	2659	1347	33.62
	Total	**4615**	**3125**	**1490**	32.29
Department of Horticulture					
1	**Group –A**	**216**	**178**	**155**	17.59
2	**Group –B**	**1526**	**857**	**818**	43.84
	Total	**1742**	**1035**	**973**	40.59
Department of Animal Husbandry & Veterinary Services					
1	**Group –A**	**3005**	**2388**	**617**	20.53
2	**Group –B**	**273**	**12**	**261**	95.60
	Total	**3278**	**2400**	**878**	26.78
Department of Fisheries					
1	**Group –A**	40	37	5	7.50
2	**Group –B**	223	157	67	29.60
	Total	**263**	**194**	**72**	26.24
Department of Sericulture					
1	**Group –A**	91	68	23	25.27
2	**Group –B**	1496	1161	335	22.39
	Total	**1587**	**1229**	**358**	22.56
Forest Department					
1	**Group –A**	156	141	15	9.62
2	**Group –B**	3546	2262	1254	36.21
	Total	**3702**	**2403**	**1269**	35.09

Women and child development					
1	Group –A	14	10	4	28.57
2	Group –B	41	30	11	26.82
	Total	**55**	**40**	**15**	**29.13**
Grand Total					
1	Group –A	4131	3288	962	20.38
2	Group –B	11111	7138	4093	35.79
	Total	**15242**	**10426**	**5055**	28.08

Annexure 4.1: Students out turn in Agriculture and Allied Sectors during 2010-11 to 2014-15 and supply projection 2020 & 2025

Annex. 4.1.1: Students out turn in Agriculture Science during 2010-11 to 2014-15 and supply projection, 2020 & 2025

Year	Diploma	UG	PG	Ph.D.	Total
2010-11	223	614	460	86	**1160**
2011-12	565	590	413	99	**1102**
2012-13	548	674	448	92	**1214**
2013-14	548	780	507	129	**1416**
2014-15	722	955	637	126	**1718**
CAGR (%)	26.10	12.33	8.94	10.83	10.92
2015-16	811	1073	694	140	1930
2016-17	911	1205	756	155	**2141**
2017-18	1023	1354	824	172	**2374**
2018-19	1149	1520	897	190	**2633**
2019-20	1291	1708	977	211	2921
2020-21	1450	1918	1065	234	**3240**
2021-22	1629	2155	1160	259	**3594**
2022-23	1830	2421	1264	287	**3986**
2023-24	2056	2719	1377	318	**4421**
2024-25	**2309**	**3054**	**1500**	**352**	**4904**

Annex. 4.1.2: Students out turn in Agriculture Markrting and Co-operation Courses during 2010-11 to 2014-15 and supply projection, 2020 & 2025

Year	UG	PG	Ph.D.	Total
2010-11	49	26	2	**77**
2011-12	61	28	3	**92**
2012-13	70	32	4	**106**
2013-14	106	30	5	**141**
2014-15	69	33	3	**105**
CAGR (%)	13.17	5.61	14.13	11.04
2015-16	78	35	3	119
2016-17	88	37	4	**132**
2017-18	100	39	4	**147**
2018-19	113	41	5	**163**
2019-20	128	43	6	181
2020-21	145	46	7	**201**
2021-22	164	48	8	**223**
2022-23	186	51	9	**247**
2023-24	210	54	10	**275**
2024-25	**238**	**57**	**11**	**305**

Annex. 4.1.3: Students out turn in Agriculture Engineering Courses during 2010-11 to 2014-15 and supply projection, 2020 & 2025

Year	Diploma	UG	PG	Ph.D.	Total
2010-11	28	69	19	1	**89**
2011-12	48	55	23	1	**79**
2012-13	68	63	23	2	**88**
2013-14	59	95	26	2	**123**
2014-15	59	74	23	1	**98**
CAGR (%)	18.49	7.11	5.18	7.18	6.56
2015-16	63	79	24	1	105
2016-17	68	85	25	1	**112**
2017-18	72	91	27	1	**119**
2018-19	78	97	28	1	**127**
2019-20	83	104	30	1	135
2020-21	89	112	31	2	**144**
2021-22	95	120	33	2	**154**
2022-23	102	128	34	2	**164**
2023-24	109	137	36	2	**175**
2024-25	**117**	**147**	**38**	**2**	**186**

Annex. 4.1.4: Students out turn in Agriculture Biotechnology Courses during 2010-11 to 2014-15 and supply projection, 2020 & 2025

Year	**UG**	**PG**	**Ph.D.**	**Total**
2010-11	37	30	8	**75**
2011-12	37	25	10	**72**
2012-13	29	44	9	**82**
2013-14	42	29	10	**81**
2014-15	46	36	9	**91**
CAGR (%)	5.78	5.26	2.38	5.17
2015-16	49	38	9	96
2016-17	51	40	9	**101**
2017-18	54	42	10	**106**
2018-19	58	44	10	**112**
2019-20	61	47	10	118
2020-21	64	49	10	**124**
2021-22	68	52	11	**130**
2022-23	72	54	11	**137**
2023-24	76	57	11	**144**

Annex.4.1.5: Students out turn in Rural Home Science Courses during 2010-11 to 2014-15 and supply projection, 2020 & 2025.

Year	UG	PG	Ph.D.	Total
2010-11	28	22	8	**58**
2011-12	33	16	7	**56**
2012-13	34	9	8	**51**
2013-14	30	21	11	**62**
2014-15	39	24	10	**73**
CAGR (%)	5.84	4.56	9.40	5.78
2015-16	41	25	11	77
2016-17	44	26	12	**82**
2017-18	46	27	13	**86**
2018-19	49	29	14	**91**
2019-20	52	30	16	97
2020-21	55	31	17	**102**
2021-22	58	33	19	**108**
2022-23	61	34	21	**114**
2023-24	65	36	22	**121**
2024-25	**69**	**37**	**25**	**128**

Annex.4.1.6: Students out turn in Food Technology Courses during 2010-11 to 2014-15 and supply projection, 2020 & 2025

Year	UG	PG	Ph.D.	Total
2010-11	31	4	1	**36**
2011-12	34	4	1	**39**
2012-13	45	4	1	**50**
2013-14	43	5	2	**50**
2014-15	38	6	2	**46**
CAGR (%)	6.63	10.89	23.11	7.67
2015-16	41	7	2	49
2016-17	43	7	3	**53**
2017-18	46	8	4	**57**
2018-19	49	9	5	**61**
2019-20	52	10	6	66
2020-21	56	11	7	**71**
2021-22	60	12	9	**76**
2022-23	64	14	11	**82**
2023-24	68	15	13	**89**
2024-25	**72**	**17**	**16**	**95**

Annex.4.1.7: Students out turn in Sericulture Courses during 2010-11 to 2014-15 and supply projection, 2020 & 2025

Year	Diploma	UG	PG	Ph.D.	Total
2010-11	9	10	6	2	**18**
2011-12	11	13	9	2	**24**
2012-13	12	12	4	2	**18**
2013-14	14	15	11	2	**28**
2014-15	15	15	12	4	**31**
CAGR (%)	13.46	10.01	17.20	14.87	13.22
2015-16	17	17	14	5	34
2016-17	18	18	16	5	**39**
2017-18	20	20	19	6	**44**

2018-19	22	22	23	7	**49**
2019-20	24	24	27	8	56
2020-21	27	27	31	9	**63**
2021-22	29	29	36	11	**72**
2022-23	32	32	43	12	**81**
2023-24	35	35	50	14	**92**
2024-25	**39**	**39**	**59**	**16**	**104**

Annex.4.1.8: Students out turn in Forestry Courses during 2010-11 to 2014-15 and supply projection, 2020 & 2025.

Year	UG	PG	Ph.D.	Total
2010-11	30	23	2	**55**
2011-12	44	28	1	**73**
2012-13	106	22	5	**133**
2013-14	82	22	6	**110**
2014-15	60	35	4	**99**
CAGR (%)	22.25	6.17	33.79	17.06
2015-16	73	37	5	120
2016-17	90	39	6	**141**
2017-18	110	42	8	**165**
2018-19	134	44	11	**193**
2019-20	164	47	15	226
2020-21	200	50	20	**265**
2021-22	245	53	27	**310**
2022-23	299	56	36	**363**
2023-24	366	60	48	**425**
2024-25	**447**	**64**	**64**	**497**

Annex.4.1.9: Students out turn in Horticulture Courses during 2010-11 to 2014-15 and supply projection, 2020 & 2025.

Year	Diploma	UG	PG	Ph.D.	Total
2010-11	22	180	75	20	**275**
2011-12	25	200	89	18	**307**
2012-13	31	300	90	21	**411**
2013-14	45	375	95	38	**508**
2014-15	50	249	120	35	**404**
CAGR (%)	24.98	13.63	10.58	20.52	13.57
2015-16	57	283	133	42	459
2016-17	65	321	147	51	**521**
2017-18	73	365	162	61	**592**
2018-19	83	415	179	74	**673**
2019-20	95	472	198	89	764
2020-21	108	536	219	107	**868**
2021-22	122	609	243	129	**985**
2022-23	139	692	268	156	**1119**
2023-24	158	786	297	188	**1271**
2024-25	**179**	**893**	**328**	**226**	**1443**

Annex.4.1.10: Students out turn in Veterinary Courses during 2010-11 to 2014-15 and supply projection, 2020 & 2025.

Year	Diploma	UG	PG	Ph.D.	Total
2010-11	17	102	49	12	**163**
2011-12	45	148	79	14	**241**
2012-13	94	162	45	12	**219**
2013-14	97	117	105	22	**244**
2014-15	97	151	77	18	**246**
CAGR (%)	52.98	5.65	12.62	13.46	8.71
2015-16	102	160	87	20	260
2016-17	108	169	98	23	**283**
2017-18	114	178	110	26	**307**
2018-19	121	188	124	30	**334**
2019-20	128	199	139	34	363
2020-21	135	210	157	38	**395**
2021-22	143	222	177	44	**429**
2022-23	151	234	199	49	**466**
2023-24	159	248	224	56	**507**
2024-25	**168**	**262**	**253**	**64**	**551**

Annex.4.1.11: Students out turn in Fishery Science Courses during 2010-11 to 2014-15 and supply projection, 2020 & 2025.

Year	UG	PG	Ph.D.	Total
2010-11	16	10	2	**28**
2011-12	14	13	1	**28**
2012-13	20	7	4	**31**
2013-14	41	9	1	**51**
2014-15	22	15	6	**43**
CAGR (%)	18.67	4.53	24.57	15.69
2015-16	26	16	7	51
2016-17	31	16	9	**59**
2017-18	37	17	12	**68**
2018-19	44	18	14	**79**
2019-20	52	19	18	91
2020-21	61	20	22	**106**
2021-22	73	20	28	**122**
2022-23	87	21	35	**142**
2023-24	103	22	43	**164**
2024-25	**122**	**23**	**54**	**189**

Annex.4.1.12: Students out turn in Dairy Technology Courses during 2010-11 to 2014-15 and supply projection, 2020 & 2025.

Year	UG	PG	Ph.D.	Total
2010-11	31	3	1	**35**
2011-12	24	24	2	**50**
2012-13	31	17	1	**49**
2013-14	40	13	1	**54**
2014-15	38	7	2	**47**
CAGR (%)	9.61	11.42	7.18	6.89
2015-16	42	8	2	52
2016-17	46	9	2	57
2017-18	50	10	2	62
2018-19	55	11	3	68
2019-20	**60**	**12**	**3**	75
2020-21	66	13	3	82
2021-22	72	15	3	90
2022-23	79	17	3	99
2023-24	87	19	4	109
2024-25	**95**	**21**	**4**	120

Annex.4.1.13: Required Annual Out turn by Education Level in 2014-15 (Supply) & 2024-25 (Demand)

Discipline	UG		PG		Ph.D.		UG and above		
	2014-15	2024-25	2014-15	2024-25	2014-15	2024-25	Supply- 2014-15	Demand- 2024-25	Gap
Agriculture	955	2679	637	1649	126	674	1718	5002	3284
Agril. Marketing & Cop	69	423	33	206	3	29	105	658	553
Agril. Engineering	74	150	23	158	1	41	98	349	251
Agril. Biotechnology	46	108	36	101	9	32	91	241	150
Rural Home Science	39	97	24	85	10	38	73	220	147
Food Technology	38	308	0	64	0	21	38	394	356
Sericulture	15	289	12	136	4	16	31	441	410
Forestry	60	951	35	446	4	38	99	1435	1336
Horticulture	249	1074	120	677	35	367	404	2118	1714
Veterinary	151	971	77	502	18	224	246	1697	1451
Fishery	22	146	15	98	6	59	43	304	261
Dairy Technology	38	189	7	98	2	29	47	316	269
Total	**1756**	**7386**	**1019**	**4220**	**218**	**1569**	**2993**	**13175**	**10183**

Annexure 4.2 Human Resources Stock Requirement Projections (demand) for Agriculture

Annex. 4.2.1: Agriculture Human Resources Stock Requirement Projections for State Governments Departments/organization (5%)

Year	Diploma	UG	PG	Ph.D.	Total
2014-15	75	1135	525	34	1694
2015-16	79	1192	551	36	1779
2016-17	83	1251	579	37	1868
2017-18	87	1314	608	39	1961
2018-19	91	1380	638	41	2059
2019-20	**96**	**1449**	**670**	**43**	**2162**
2020-21	101	1521	704	45	2270
2021-22	106	1597	739	48	2384
2022-23	111	1677	776	50	2503
2023-24	116	1761	815	53	2628
2024-25	**122**	**1849**	**855**	**55**	**2759**

Annex.4.2.2: Agriculture Human Resources Stock Requirement Projections for State Agriculture Universities (Teaching, Research and Extension) (5%)

Year	Diploma	UG	PG	Ph.D.	Total
2014-15	55	80	106	346	532
2015-16	58	84	112	363	559
2016-17	61	88	117	381	587
2017-18	64	92	123	400	616
2018-19	67	97	129	420	647
2019-20	70	102	136	441	679
2020-21	74	107	143	463	713
2021-22	77	112	150	487	749
2022-23	81	118	157	511	786
2023-24	85	124	165	536	825
2024-25	**90**	**130**	**173**	**563**	**867**

Annex.4.2.3: Agriculture Human Resources Stock Requirement Projections for Financial Institutions (5.6 %)

Year	UG	PG	Ph.D.	Total
2014-15	193	158	18	350
2015-16	203	166	18	370
2016-17	215	176	20	390
2017-18	227	185	21	412
2018-19	239	196	22	435
2019-20	**253**	**207**	**23**	**460**
2020-21	267	218	24	485
2021-22	282	231	26	513
2022-23	298	244	27	541
2023-24	314	257	29	572
2024-25	**332**	**272**	**30**	**604**

Annex.4.2.4: Agriculture Human Resources Stock Requirement Projections for Seed Sector (10%)

Year	Diploma	UG	PG	Ph.D.	Total
2014-15	85	32	41	5	90
2015-16	89	35	45	5	99
2016-17	94	38	49	5	109
2017-18	98	42	54	6	120
2018-19	103	46	59	7	132
2019-20	108	51	65	7	145
2020-21	114	56	72	8	159
2021-22	120	61	79	9	175
2022-23	126	68	87	10	193
2023-24	132	74	95	11	212
2024-25	**138**	**82**	**105**	**12**	**233**

Annex.4.2.5: Agriculture Human Resources Stock Requirement Projections for Fertilizer Industries (5%)

Year	Diploma	UG	PG	Ph.D.	Total
2014-15	80	25	32	4	70
2015-16	84	29	26	4	74
2016-17	88	31	27	4	77
2017-18	93	32	28	4	81
2018-19	97	34	30	4	85
2019-20	**102**	**36**	**31**	**4**	**89**
2020-21	107	38	33	5	94
2021-22	113	39	34	5	98
2022-23	118	41	36	5	103
2023-24	124	43	38	5	109
2024-25	**130**	**46**	**40**	**6**	**114**

Annex.4.2.6: Agriculture Human Resources Stock Requirement Projections for Pesticide Industries (3%)

Year	Diploma	UG	PG	Ph.D.	Total
2014-15	65	18	23	3	50
2015-16	68	18	21	3	52
2016-17	72	19	21	3	53
2017-18	75	19	22	3	55
2018-19	79	20	23	3	56
2019-20	83	20	23	3	58
2020-21	87	21	24	3	60
2021-22	91	22	25	3	61
2022-23	96	22	25	3	63
2023-24	101	23	26	3	65
2024-25	**106**	**24**	**27**	**3**	**67**

Annex.4.2.7: Agriculture Human Resources Stock Requirement Projections for NGO's (3%)

Year	Diploma	UG	PG	Ph.D.	Total
2014-15	8	18	23	3	50
2015-16	8	23	20	1	52
2016-17	8	24	21	1	53
2017-18	8	25	21	1	55
2018-19	8	25	22	1	56
2019-20	**9**	**26**	**23**	**1**	**58**
2020-21	9	27	23	1	60
2021-22	9	28	24	1	61
2022-23	10	29	25	1	63
2023-24	10	29	25	1	65
2024-25	**10**	**30**	**26**	**1**	**67**

Annex.4.2.8: Agriculture Human Resources Stock Requirement Projections for Agro Processing Sector (5%)

Year	Diploma	UG	PG	Ph.D.	Total
2014-15	8	18	23	3	50
2015-16	8	26	15	1	53
2016-17	8	28	16	1	55
2017-18	9	29	17	1	58
2018-19	9	30	18	1	61
2019-20	10	32	19	1	64
2020-21	10	34	19	1	67
2021-22	11	35	20	1	70
2022-23	11	37	21	1	74
2023-24	12	39	22	1	78
2024-25	**12**	**41**	**24**	**1**	**81**

Annex.4.2.9: Agriculture Human Resources Stock Requirement Projections for Other Sectors (5%)

Year	Diploma	UG	PG	Ph.D.	Total
2014-15	30	70	90	10	200
2015-16	32	95	82	2	210
2016-17	33	99	86	2	221
2017-18	35	104	90	2	232
2018-19	36	109	95	2	243
2019-20	**38**	**115**	**100**	**3**	**255**
2020-21	40	121	105	3	268
2021-22	42	127	110	3	281
2022-23	44	133	115	3	295
2023-24	47	140	121	3	310
2024-25	**49**	**147**	**127**	**3**	**326**

Annex.4.2.10: Agriculture Human Resources Stock Requirement Projections for Agriculture Sector (Total)

Year	Diploma	UG	PG	Ph.D.	Total
2014-15	820	1586	1019	423	3086
2015-16	861	1705	1038	431	3245
2016-17	904	1792	1092	453	3412
2017-18	949	1884	1149	476	3589
2018-19	997	1981	1209	501	3774
2019-20	1047	2083	1273	526	3970
2020-21	1099	2190	1340	553	4176
2021-22	1154	2303	1411	581	4393
2022-23	1212	2422	1486	610	4622
2023-24	1272	2547	1565	641	4864
2024-25	**1336**	**2679**	**1649**	**674**	**5119**

Annex. 4.2.11 Demand Supply Gap in Agriculture Sector

Year	Diploma	UG	PG	Ph.D.	Total
2015-16	**50**	**632**	**344**	**292**	**1315**
2016-17	93	587	336	299	1272
2017-18	138	531	326	305	1214
2018-19	186	460	312	310	1141
2019-20	**236**	**375**	**296**	**315**	**1049**
2020-21	288	272	276	319	936
2021-22	343	148	252	322	800
2022-23	400	1	223	323	637
2023-24	461	-172	189	323	443
2024-25	**525**	**-375**	**149**	**322**	**215**

Annex. 4.2.12 Total Demand Projected (Low Growth)

Year	Diploma	UG	PG	Ph.D.	Total
2014-15	820	1586	1019	423	3086
2015-16	845	1633	1049	435	3179
2016-17	870	1682	1081	448	3274
2017-18	896	1733	1113	462	3372
2018-19	923	1785	1146	476	3473
2019-20	951	1838	1181	490	3578
2020-21	979	1894	1216	505	3685
2021-22	1008	1950	1253	520	3795
2022-23	1039	2009	1290	535	3909
2023-24	1070	2069	1329	552	4027
2024-25	**1102**	**2131**	**1369**	**568**	**4147**

Annex. 4.2.13: Total Demand Projected (Average Growth)

Year	Diploma	UG	PG	Ph.D.	Total
2014-15	820	1586	1019	423	3086
2015-16	861	1665	1069	444	3240
2016-17	904	1748	1123	466	3402
2017-18	949	1836	1179	489	3572
2018-19	997	1928	1238	514	3751
2019-20	**1047**	**2024**	**1300**	**539**	**3939**
2020-21	1099	2125	1365	566	4136
2021-22	1154	2231	1433	595	4342
2022-23	1212	2343	1505	624	4559
2023-24	1272	2460	1580	656	4787
2024-25	**1336**	**2583**	**1659**	**689**	**5027**

Annex. 4.2.14: Agriculture Human Resources Stock Requirement Projections for Agriculture Sector (High Growth)

Year	Diploma	UG	PG	Ph.D.	Total
2014-15	820	1586	1019	423	3086
2015-16	886	1713	1100	456	3333
2016-17	956	1850	1188	493	3600
2017-18	1033	1998	1283	532	3887
2018-19	1116	2157	1386	575	4198
2019-20	1205	2330	1497	621	4534
2020-21	1301	2516	1616	671	4897
2021-22	1405	2718	1746	724	5289
2022-23	1518	2935	1885	782	5712
2023-24	1639	3170	2036	845	6169
2024-25	1770	3424	2199	913	6662

Annexure 4.3 Human Resources Stock Requirement Projections (demand) for Agriculture Marketing & Co-operation

Annex. 4.3.1: Agriculture marketing and co-operation Human Resources Stock Requirement Projections for Agriculture Marketing Department Govt. of Karnataka (5%)

Year	UG	PG	Ph.D.	Total
2014-15	208	96	6	310
2015-16	218	101	7	326
2016-17	229	106	7	342
2017-18	240	111	7	359
2018-19	252	117	8	377
2019-20	**265**	**123**	**8**	**396**
2020-21	278	129	8	415
2021-22	292	135	9	436
2022-23	307	142	9	458
2023-24	322	149	10	481
2024-25	**338**	**157**	**10**	**505**

Annex. 4.3.2: Agriculture marketing and co-operation Human Resources Stock Requirement Projections for State Agriculture Universities (Teaching, Research and Extension) (5%)

Year	UG	PG	Ph.D.	Total
2014-15	2	2	7	10
2015-16	2	2	7	11
2016-17	2	2	7	11
2017-18	2	2	8	12
2018-19	2	2	8	12
2019-20	2	3	8	13
2020-21	9	4	9	13
2021-22	9	4	9	14
2022-23	10	5	10	15
2023-24	10	5	10	16
2024-25	**11**	**5**	**11**	**16**

Annex. 4.3.3: Agriculture marketing and co-operation Human Resources Stock Requirement Projections for Financial Institutions (5.6 %)

Year	UG	PG	Ph.D.	Total
2014-15	11	14	1	25
2015-16	12	15	1	26
2016-17	13	15	1	28
2017-18	13	16	1	29
2018-19	14	17	2	31
2019-20	**15**	**18**	**2**	**33**
2020-21	23	11	2	34
2021-22	24	11	2	36
2022-23	25	12	2	38
2023-24	27	12	2	40
2024-25	**28**	**13**	**2**	**42**

Annex. 4.3.4: Agriculture marketing and co-operation Human Resources Stock Requirement Projections for Seed, Fertilizer and Pesticide sector (8%)

Year	UG	PG	Ph.D.	Total
2014-15	6	8	2	15
2015-16	6	8	2	16
2016-17	7	9	2	17
2017-18	8	9	2	19
2018-19	8	10	2	20
2019-20	9	11	2	22
2020-21	16	12	2	23
2021-22	16	12	2	24
2022-23	17	13	3	26
2023-24	18	13	3	27
2024-25	**19**	**14**	**3**	**28**

Annex. 4.3.5: Agriculture marketing and co-operation Human Resources Stock Requirement Projections for Agro Processing Sector (5%)

Year	UG	PG	Ph.D.	Total
2014-15	6	4	1	10
2015-16	6	4	1	11
2016-17	6	4	1	11
2017-18	6	5	1	12
2018-19	7	5	1	12
2019-20	**7**	**5**	**1**	**13**
2020-21	7	5	1	13
2021-22	8	6	1	14
2022-23	8	6	1	15
2023-24	9	6	1	16
2024-25	**11**	**7**	**1**	**16**

Annex. 4.3.6: Agriculture marketing and co-operation Human Resources Stock Requirement Projections for Other Sectors (5%)

Year	UG	PG	Ph.D.	Total
2014-15	7	7	2	15
2015-16	7	7	2	16
2016-17	7	7	2	17
2017-18	8	8	2	17
2018-19	8	8	2	18
2019-20	**9**	**9**	**2**	**19**
2020-21	9	9	2	20
2021-22	9	9	2	21
2022-23	10	10	2	22
2023-24	10	10	2	23
2024-25	**16**	**11**	**2**	**24**

Annex. 4.3.7: Agriculture marketing and co-operation Human Resources Stock Requirement Projections (Total)

Year	UG	PG	Ph.D.	Total
2014-15	**239**	**130**	**17**	**385**
2015-16	251	137	18	405
2016-17	264	144	19	426
2017-18	277	152	20	448
2018-19	291	160	21	471
2019-20	**306**	**168**	**23**	**495**
2020-21	342	170	24	520
2021-22	359	178	25	546
2022-23	377	187	26	573
2023-24	396	196	27	602
2024-25	**423**	**206**	**29**	**632**

Annex. 4.3.8: Demand Supply Gap in Agriculture marketing and co-operation Sector

Year	UG	PG	Ph.D.	Total
2015-16	173	102	15	286
2016-17	175	107	15	294
2017-18	177	113	16	301
2018-19	178	119	16	308
2019-20	**178**	**125**	**17**	**315**
2020-21	197	124	17	319
2021-22	195	130	17	323
2022-23	192	136	18	326
2023-24	186	142	18	327
2024-25	**186**	**149**	**18**	**327**

Annex. 4.3.9: Human Resources Stock Requirement Projections for Agriculture marketing and co-operation Sector (Low Growth)

Year	UG	PG	Ph.D.	Total
2014-15	**239**	**130**	**17**	**385**
2015-16	246	134	18	397
2016-17	253	138	19	408
2017-18	261	142	19	421
2018-19	269	146	20	433
2019-20	**277**	**151**	**20**	**446**
2020-21	285	155	21	460
2021-22	294	160	21	474
2022-23	302	165	22	488
2023-24	311	170	23	502
2024-25	**321**	**175**	**23**	**517**

Annex. 4.3.10: Human Resources Stock Requirement Projections for Agriculture marketing and co-operation Sector (Average Growth)

Year	UG	PG	Ph.D.	Total
2014-15	239	130	17	385
2015-16	251	137	18	404
2016-17	263	143	19	424
2017-18	276	151	20	446
2018-19	290	158	21	468
2019-20	**305**	**166**	**22**	**491**
2020-21	320	174	23	516
2021-22	336	183	25	542
2022-23	353	192	26	569
2023-24	370	202	27	597
2024-25	**389**	**212**	**28**	**627**

Annex. 4.3.11: Human Resources Stock Requirement Projections for Agriculture marketing and co-operation Sector (High Growth)

Year	UG	PG	Ph.D.	Total
2014-15	**239**	**130**	**17**	**385**
2015-16	258	141	19	416
2016-17	278	152	20	449
2017-18	301	164	22	485
2018-19	325	177	24	524
2019-20	**351**	**191**	**26**	**566**
2020-21	379	206	28	611
2021-22	409	223	30	660
2022-23	442	241	32	713
2023-24	477	260	35	770
2024-25	**515**	**281**	**38**	**831**

Annexure 4.4: Human Resources Stock Requirement Projections (demand) for Agriculture Engineering

Annx.4.4.1: Agriculture Engineering Human Resources Stock Requirement Projections for Development Department (5%)

Year	Diploma	UG	PG	Ph.D.	Total
2014-15	**5**	**17**	**8**	**1**	**25**
2015-16	5	18	8	1	26
2016-17	6	18	9	1	28
2017-18	6	19	9	1	29
2018-19	6	20	9	1	30
2019-20	**6**	**21**	**10**	**1**	**32**
2020-21	7	22	10	1	34
2021-22	7	24	11	1	35
2022-23	7	25	11	1	37
2023-24	8	26	12	1	39
2024-25	**8**	**27**	**13**	**1**	**41**

Annx.4.4.2: Agriculture Engineering Human Resources Stock Requirement Projections for SAUs (5%)

Year	Diploma	UG	PG	Ph.D.	Total
2014-15	**10**	**2**	**3**	**10**	**15**
2015-16	11	2	3	10	16
2016-17	11	2	3	11	17
2017-18	12	3	3	11	17
2018-19	12	3	4	12	18
2019-20	**13**	**3**	**4**	**12**	**19**
2020-21	13	**3**	**4**	**13**	20
2021-22	14	**3**	**4**	**14**	21
2022-23	15	**3**	**4**	**14**	22
2023-24	16	**3**	**5**	**15**	23
2024-25	**16**	**4**	**5**	**16**	**24**

Annx. 4.4.3: Agriculture Engineering Human Resources Stock Requirement Projections for Financial institutions (5.6%)

Year	Diploma	UG	PG	Ph.D.	Total
2014-15	**5**	**2**	**3**	**0**	**5**
2015-16	5	2	3	0	5
2016-17	6	3	3	0	6
2017-18	6	3	3	0	6
2018-19	6	3	3	0	6
2019-20	7	**3**	**4**	**0**	7
2020-21	7	5	2	0	7
2021-22	7	5	2	0	7
2022-23	8	5	2	0	8
2023-24	8	5	3	0	8
2024-25	9	**6**	**3**	**0**	9

Annx. 4.4.5: Agriculture Engineering Human Resources Stock Requirement Projections for farm machinery sector (10%)

Year	Diploma	UG	PG	Ph.D.	Total
2014-15	**15**	**38**	**48**	**10**	**95**
2015-16	17	42	52	10	105
2016-17	18	46	57	11	115
2017-18	20	51	63	13	126
2018-19	22	56	70	14	139
2019-20	**24**	**61**	**76**	**15**	**153**
2020-21	27	67	84	17	168
2021-22	29	74	93	19	185
2022-23	32	81	102	20	204
2023-24	35	90	112	22	224
2024-25	**39**	**99**	**123**	**25**	**246**

Annx. 4.4.6: Agriculture Engineering Human Resources Stock Requirement Projections for NGO (3%)

Year	Diploma	UG	PG	Ph.D.	Total
2014-15	**10**	**6**	**4**	**1**	**10**
2015-16	10	6	4	1	10
2016-17	11	6	4	1	11
2017-18	11	6	4	1	11
2018-19	11	6	5	1	11
2019-20	**12**	**6**	**5**	**1**	**12**
2020-21	12	7	5	1	12
2021-22	12	7	5	1	12
2022-23	13	7	5	1	13
2023-24	13	7	5	1	13
2024-25	**13**	**7**	**5**	**1**	**13**

Annx. 4.4.7: Agriculture Engineering Human Resources Stock Requirement Projections for other sector (2%)

Year	Diploma	UG	PG	Ph.D.	Total
2014-15	**15**	**11**	**11**	**3**	**25**
2015-16	15	11	11	3	26
2016-17	16	12	12	3	26
2017-18	16	12	12	3	27
2018-19	16	12	12	3	27
2019-20	**17**	**12**	**12**	**3**	**28**
2020-21	17	13	13	3	28
2021-22	17	13	13	3	29
2022-23	18	13	13	3	29
2023-24	18	13	13	3	30
2024-25	**18**	**14**	**14**	**3**	**30**

Annx.4.4.8: Total Demand Projected for Agriculture Engineering Human Resources Stock Requirement

Year	Diploma	UG	PG	Ph.D.	Total
2014-15	**55**	**73**	**73**	**20**	**166**
2015-16	58	78	78	22	178
2016-17	61	84	84	23	191
2017-18	64	90	91	25	206
2018-19	68	96	98	26	221
2019-20	**71**	**103**	**106**	**28**	**238**
2020-21	76	111	115	30	257
2021-22	80	120	124	33	277
2022-23	85	129	135	35	299
2023-24	90	139	146	38	323
2024-25	**95**	**150**	**158**	**41**	**349**

Annx4.4.9: Projection of Demand Supply Gap of Agriculture Engineering Human Resources Stock Requirement

Year	Diploma	UG	PG	Ph.D.	Total
2015-16	**-8**	**10**	**10**	**-43**	**103**
2016-17	**-5**	-1	59	22	**80**
2017-18	**-2**	-1	64	23	**87**
2018-19	**1**	-1	70	25	**94**
2019-20	**4**	**-1**	**77**	**27**	**103**
2020-21	**8**	-1	84	29	**112**
2021-22	**12**	0	92	31	**123**
2022-23	**17**	1	100	33	**134**
2023-24	**21**	2	110	36	**147**
2024-25	**26**	**3**	**120**	**39**	**162**

Annx.4.4.10: Total Demand Projected for of Agriculture Engineering Human Resources (Low Growth)

Year	Diploma	UG	PG	Ph.D.	Total
2014-15	**25**	**73**	**73**	**20**	**166**
2015-16	26	75	75	21	171
2016-17	26	78	77	21	176
2017-18	27	80	79	22	181
2018-19	27	82	82	23	187
2019-20	**28**	**85**	**84**	**23**	**192**
2020-21	28	87	87	24	198
2021-22	29	90	89	25	204
2022-23	29	93	92	26	210
2023-24	30	95	95	26	217
2024-25	**30**	**98**	**98**	**27**	**223**

Annx.4.4.11: Total Demand Projected for of Agriculture Engineering Human Resources (Average Growth)

Year	Diploma	UG	PG	Ph.D.	Total
2014-15	**25**	**73**	**73**	**20**	**166**
2015-16	26	77	76	21	174
2016-17	26	81	80	22	183
2017-18	27	85	84	23	192
2018-19	27	89	88	24	202
2019-20	**28**	**93**	**93**	**26**	**212**
2020-21	28	98	97	27	222
2021-22	29	103	102	28	234
2022-23	29	108	107	30	245
2023-24	30	113	113	31	258
2024-25	**30**	**119**	**118**	**33**	**270**

Annx. 4.4.12: Total Demand Projected for of Agriculture Engineering Human Resources Demand Projected (High Growth)

Year	Diploma	UG	PG	Ph.D.	Total
2014-15	**25**	**73**	**73**	**20**	**166**
2015-16	26	79	79	22	179
2016-17	26	85	85	24	194
2017-18	27	92	92	25	209
2018-19	27	100	99	27	226
2019-20	**28**	**107**	**107**	**30**	**244**
2020-21	28	116	115	32	263
2021-22	29	125	125	35	284
2022-23	29	135	135	37	307
2023-24	30	146	145	40	332
2024-25	**30**	**158**	**157**	**44**	**358**

Annexure 4.5: Agriculture Biotechnology Human Resources Stock Requirement Projections

Annx. 4.5.1.Agriculture Biotechnology Human Resources Stock Requirement Projections for Government Departments (5%)

Year	UG	PG	Ph.D.	Total
2014-15	**10**	**5**	**0**	**15**
2015-16	11	5	0	16
2016-17	11	5	0	17
2017-18	12	5	0	17
2018-19	12	6	0	18
2019-20	**13**	**6**	**0**	**19**
2020-21	13	6	0	20
2021-22	14	7	0	21
2022-23	15	7	0	22
2023-24	16	7	0	23
2024-25	**16**	**8**	**0**	24

Annx.4.5.2: Agriculture Biotechnology Human Resources Stock Requirement Projections for State Agriculture Universities (Teaching, Research and Extension) (10%)

Year	UG	PG	Ph.D.	Total
2014-15	**0**	**1**	**4**	**5**
2015-16	0	1	5	6
2016-17	0	1	5	6
2017-18	0	1	6	7
2018-19	0	1	6	7
2019-20	**0**	**1**	**7**	**8**
2020-21	0	1	8	9
2021-22	0	1	8	10
2022-23	1	1	9	11
2023-24	1	1	10	12
2024-25	**1**	**1**	**11**	**13**

Annx.4.5.3: Agriculture Biotechnology Human Resources Stock Requirement Projections for Tissue culture labs (12%)

Year	UG	PG	Ph.D.	Total
2014-15	**10**	**13**	**3**	**25**
2015-16	11	14	3	28
2016-17	13	16	3	31
2017-18	14	18	4	35
2018-19	16	20	4	39
2019-20	**18**	**22**	**4**	**44**
2020-21	**20**	**25**	**5**	49
2021-22	**22**	**28**	**6**	55
2022-23	**25**	**31**	**6**	62
2023-24	**28**	**35**	**7**	69
2024-25	**31**	**39**	**8**	**78**

Annx.4.5.4: Agriculture Biotechnology Human Resources Stock Requirement Projections for Agro Chemicals, Pharmaceuticals & Vaccination Lab, *etc.* (12%)

Year	UG	PG	Ph.D.	Total
2014-15	**5**	**6**	**1**	**15**
2015-16	6	7	1	17
2016-17	7	8	1	19
2017-18	7	8	1	21
2018-19	8	9	1	24
2019-20	**9**	**11**	**1**	**26**
2020-21	10	12	1	30
2021-22	12	13	2	33
2022-23	13	15	2	37
2023-24	15	17	2	42
2024-25	**16**	**19**	**2**	**47**

Annx.4.5.5: Agriculture Biotechnology Human Resources Stock Requirement Projections for Seed industry (12%)

Year	UG	PG	Ph.D.	Total
2014-15	**6**	**7**	**3**	**16**
2015-16	6	8	4	18
2016-17	7	9	4	20
2017-18	8	10	4	22
2018-19	9	11	5	25
2019-20	**10**	**13**	**6**	**28**
2020-21	11	14	6	32
2021-22	12	16	7	35
2022-23	14	18	8	40
2023-24	16	20	9	44
2024-25	**17**	**22**	**10**	**50**

Annx.4.5.6: Agriculture Biotechnology Human Resources Stock Requirement Projections for Other Sectors (7%)

Year	UG	PG	Ph.D.	Total
2014-15	**13**	**6**	**1**	**20**
2015-16	14	7	1	21
2016-17	15	7	1	23
2017-18	16	8	1	25
2018-19	17	8	1	26
2019-20	**18**	**9**	**1**	**28**
2020-21	20	9	1	30
2021-22	22	10	1	32
2022-23	23	11	1	34
2023-24	25	11	1	37
2024-25	**26**	**12**	**1**	**39**

Annx.4.5.7: Agriculture Biotechnology Human Resources Stock Requirement Projections (Total)

Year	UG	PG	Ph.D.	Total
2014-15	**44**	**37**	**12**	**96**
2015-16	48	41	13	105
2016-17	52	45	14	116
2017-18	57	50	16	127
2018-19	62	55	18	140
2019-20	**68**	**61**	**20**	**154**
2020-21	75	67	21	170
2021-22	82	74	24	187
2022-23	90	82	26	206
2023-24	99	91	29	227
2024-25	**108**	**101**	**32**	**251**

Annx. 4.5.8: Demand Supply Gap in Agriculture Biotechnology Sector

Year	UG	PG	Ph.D.	Total
2015-16	**1**	**3**	**5**	**7**
2016-17	4	7	6	17
2017-18	9	12	8	28
2018-19	14	17	10	41
2019-20	**20**	**23**	**12**	**54**
2020-21	38	18	1	57
2021-22	40	18	1	60
2022-23	42	19	1	63
2023-24	44	20	1	66
2024-25	**46**	**21**	**1**	**69**

Annx.4.5.9: Human Resources Stock Requirement Projections for Agriculture Biotechnology Sector (Low Growth)

Year	UG	PG	Ph.D.	Total
2014-15	**44**	**37**	**12**	**96**
2015-16	45	38	12	99
2016-17	47	39	13	102
2017-18	48	40	13	105
2018-19	50	42	13	108
2019-20	**51**	**43**	**14**	**111**
2020-21	53	44	14	115
2021-22	54	46	15	118
2022-23	56	47	15	122
2023-24	58	48	15	125
2024-25	**59**	**50**	**16**	**129**

Annx.4.5.10: Human Resources Stock Requirement Projections for Agriculture Biotechnology Sector (Average Growth)

Year	UG	PG	Ph.D.	Total
2014-15	**44**	**37**	**12**	**96**
2015-16	46	39	12	101
2016-17	49	41	13	106
2017-18	51	43	14	111
2018-19	54	45	14	117
2019-20	**56**	**47**	**15**	**123**
2020-21	59	50	16	129
2021-22	62	52	17	135
2022-23	65	55	17	142
2023-24	68	57	18	149
2024-25	**72**	**60**	**19**	**156**

Annx.4.5.11: Human Resources Stock Requirement Projections for Agriculture Biotechnology Sector (High Growth)

Year	UG	PG	Ph.D.	Total
2014-15	**44**	**37**	**12**	**96**
2015-16	48	40	13	104
2016-17	51	43	14	112
2017-18	56	47	15	121
2018-19	60	50	16	131
2019-20	**65**	**54**	**17**	**141**
2020-21	70	59	19	152
2021-22	76	63	20	165
2022-23	82	69	22	178
2023-24	88	74	24	192
2024-25	**95**	**80**	**25**	**207**

Annexure 4.6: Human Resources Stock Requirement Projections of Rural Home Science

Annx.4.6.1: Rural Home Science Human Resources Stock Requirement Projections for Government Departments (5%)

Year	UG	PG	Ph.D.	Total
2014-15	**10**	**5**	**0**	**15**
2015-16	11	5	0	17
2016-17	12	6	0	18
2017-18	13	6	0	20
2018-19	15	7	0	22
2019-20	**16**	**7**	**0**	24
2020-21	18	8	1	27
2021-22	20	9	1	29
2022-23	22	10	1	32
2023-24	24	11	1	35
2024-25	**26**	**12**	**1**	39

Annx.4.6.2: Rural Home Science Human Resources Stock Requirement Projections for State Agriculture Universities (Teaching, Research and Extension) (10%)

Year	UG	PG	Ph.D.	Total
2014-15	**2**	**3**	**10**	**15**
2015-16	2	3	11	17
2016-17	3	4	12	18
2017-18	3	4	13	20
2018-19	3	4	14	22
2019-20	**4**	**5**	**16**	**24**
2020-21	4	5	17	27
2021-22	4	6	19	29
2022-23	5	6	21	32
2023-24	5	7	23	35
2024-25	**6**	**8**	**25**	**39**

Annx.4.6.3: Rural Home Science Human Resources Stock Requirement Projections for Food processing and value addition industries (15%)

Year	UG	PG	Ph.D.	Total
2014-15	**10**	**13**	**3**	**25**
2015-16	12	14	3	29
2016-17	13	17	3	33
2017-18	15	19	4	38
2018-19	17	22	4	44
2019-20	**20**	**25**	**5**	**50**
2020-21	23	29	6	58
2021-22	27	33	7	67
2022-23	31	38	8	76
2023-24	35	44	9	88
2024-25	**40**	**51**	**10**	**101**

Annx.4.6.4: Rural Home Science Human Resources Stock Requirement Projections for Hospitals (2%)

Year	UG	PG	Ph.D.	Total
2014-15	**2**	**2**	**0**	**5**
2015-16	2	2	0	5
2016-17	2	2	0	5
2017-18	2	2	0	5
2018-19	2	2	0	5
2019-20	**2**	**2**	**0**	**6**
2020-21	2	2	0	6
2021-22	2	2	0	6
2022-23	2	2	0	6
2023-24	2	2	0	6
2024-25	**2**	**2**	**0**	**6**

Annx. 4.6.5: Rural Home Science Human Resources Stock Requirement Projections for self employment (2 %)

Year	UG	PG	Ph.D.	Total
2014-15	1	2	0	4
2015-16	1	2	0	4
2016-17	1	2	0	4
2017-18	1	2	0	4
2018-19	2	2	0	4
2019-20	2	2	0	4
2020-21	2	2	0	5
2021-22	2	2	0	5
2022-23	2	2	0	5
2023-24	2	2	0	5
2024-25	2	2	0	5

Annx. 4.6.6: Rural Home Science Human Resources Stock Requirement Projections for NGO's (2%)

Year	UG	PG	Ph.D.	Total
2014-15	7	3	0	10
2015-16	7	3	0	10
2016-17	7	3	0	10
2017-18	7	3	0	11
2018-19	7	3	0	11
2019-20	7	3	0	11
2020-21	7	3	0	11
2021-22	7	4	0	11
2022-23	8	4	0	12
2023-24	8	4	0	12
2024-25	8	4	0	12

Annx.4.6.7: Rural Home Science Human Resources Stock Requirement Projections for Other Sectors (3%)

Year	UG	PG	Ph.D.	Total
2014-15	**10**	**5**	**1**	**15**
2015-16	10	5	1	15
2016-17	10	5	1	16
2017-18	11	5	1	16
2018-19	11	5	1	17
2019-20	11	5	1	17
2020-21	12	6	1	18
2021-22	12	6	1	18
2022-23	12	6	1	19
2023-24	13	6	1	20
2024-25	13	6	1	20

Annx.4.6.8: Rural Home Science Human Resources Stock Requirement Projections (Total)

Year	UG	PG	Ph.D.	Total
2014-15	**42**	**32**	**14**	**89**
2015-16	45	34	15	97
2016-17	48	38	17	105
2017-18	52	41	19	115
2018-19	57	46	21	125
2019-20	**62**	**50**	**23**	**137**
2020-21	67	56	25	150
2021-22	74	62	28	165
2022-23	81	68	31	182
2023-24	88	76	34	201
2024-25	**97**	**85**	**38**	**222**

Annx.4.6.9: Demand Supply Gap in Rural Home Science Sector

Year	UG	PG	Ph.D.	Total
2015-16	**4**	**9**	**4**	**19**
2016-17	5	11	5	23
2017-18	6	14	6	28
2018-19	8	17	6	34
2019-20	**10**	**20**	**7**	**40**
2020-21	13	24	8	48
2021-22	16	29	9	57
2022-23	19	34	10	68
2023-24	23	40	12	80
2024-25	**28**	**47**	**13**	**94**

Annx.4.6.10: Human Resources Stock Requirement Projections for Rural Home Science Sector (Low Growth)

Year	UG	PG	Ph.D.	Total
2014-15	**42**	**32**	**14**	**89**
2015-16	43	32	14	92
2016-17	44	33	15	94
2017-18	46	34	15	97
2018-19	47	35	16	100
2019-20	**48**	**37**	**16**	**103**
2020-21	50	38	17	106
2021-22	51	39	17	109
2022-23	53	40	18	113
2023-24	54	41	18	116
2024-25	**56**	**42**	**19**	**120**

Annx. 4.6.11: Human Resources Stock Requirement Projections for Rural Home Science Sector (Average Growth)

Year	UG	PG	Ph.D.	Total
2014-15	**42**	**32**	**14**	**89**
2015-16	44	33	15	93
2016-17	46	35	15	98
2017-18	48	36	16	103
2018-19	51	38	17	108
2019-20	**53**	**40**	**18**	**114**
2020-21	56	42	19	119
2021-22	59	44	20	125
2022-23	62	47	21	131
2023-24	65	49	22	138
2024-25	**68**	**51**	**23**	**145**

Annx. 4.6.12: Human Resources Stock Requirement Projections for Rural Home Science Sector (High Growth)

Year	UG	PG	Ph.D.	Total
2014-15	**42**	**32**	**14**	**89**
2015-16	45	34	15	96
2016-17	49	37	16	104
2017-18	53	40	18	112
2018-19	57	43	19	121
2019-20	**61**	**46**	**21**	**131**
2020-21	66	50	22	141
2021-22	71	54	24	153
2022-23	77	58	26	165
2023-24	83	63	28	178
2024-25	**90**	**68**	**30**	**192**

Annexure 4.7: Human Resources Stock Requirement Projections for Food Technology

Annx.4.7.1: Food Technology Human Resources Stock Requirement Projections for SAUs (2%)

Year	UG	PG	Ph.D.	Total
2014-15	**6**	**6**	**13**	**25**
2015-16	6	6	13	26
2016-17	7	7	13	26
2017-18	7	7	13	27
2018-19	7	7	14	27
2019-20	7	7	14	28
2020-21	7	7	14	28
2021-22	7	7	14	29
2022-23	7	7	15	29
2023-24	7	7	15	30
2024-25	8	8	15	30

Annx.4.7.2: Food Technology Human Resources Stock Requirement Projections for Food Processing Industries (30%)

Year	UG	PG	Ph.D.	Total
2014-15	**20**	**4**	**1**	**25**
2015-16	26	5	2	33
2016-17	34	6	2	42
2017-18	44	8	3	55
2018-19	57	11	4	71
2019-20	74	14	5	93
2020-21	97	18	6	121
2021-22	125	24	8	157
2022-23	163	31	10	204
2023-24	212	40	13	265
2024-25	276	52	17	345

Annx.4.7.3: Food Technology Human Resources Stock Requirement Projections for Self Employment (2%)

Year	UG	PG	Ph.D.	Total
2014-15	**2**	**1**	**0**	**2**
2015-16	2	1	0	2
2016-17	2	1	0	2
2017-18	2	1	0	2
2018-19	2	1	0	2
2019-20	**2**	**1**	**0**	2
2020-21	2	1	0	2
2021-22	2	1	0	2
2022-23	2	1	0	2
2023-24	2	1	0	2
2024-25	**2**	**1**	**0**	2

Annx.4.7.4: Food Technology Human Resources Stock Requirement Projections for NGOs (3%)

Year	UG	PG	Ph.D.	Total
2014-15	**3**	**1**	**0**	**4**
2015-16	3	1	0	4
2016-17	3	1	0	4
2017-18	3	1	0	4
2018-19	3	1	0	5
2019-20	**3**	**1**	**0**	**5**
2020-21	4	1	0	5
2021-22	4	1	0	5
2022-23	4	1	0	5
2023-24	4	1	0	5
2024-25	**4**	**1**	**0**	**5**

Annx.4.7.5: Food Technology Human Resources Stock Requirement Projections for Other sectors (5%)

Year	UG	PG	Ph.D.	Total
2014-15	**8**	**3**	**0**	**10**
2015-16	8	3	0	11
2016-17	8	3	0	11
2017-18	9	3	0	12
2018-19	9	3	0	12
2019-20	**10**	**3**	**0**	**13**
2020-21	10	3	0	13
2021-22	11	4	0	14
2022-23	11	4	0	15
2023-24	12	4	0	16
2024-25	**12**	**4**	**0**	**16**

Annx.4.7.6: Total Demand Projected for Food Technology Human Resources Stock Requirement

Year	UG	PG	Ph.D.	Total
2014-15	**38**	**14**	**14**	**66**
2015-16	45	15	14	75
2016-17	53	17	15	86
2017-18	64	19	16	100
2018-19	78	22	17	117
2019-20	**96**	**26**	**18**	**140**
2020-21	119	30	20	169
2021-22	149	36	22	207
2022-23	187	43	25	255
2023-24	237	53	28	318
2024-25	**301**	**65**	**32**	**399**

Annx.4.7.7: Projection of Demand Supply Gap of Food Technology Human Resources

Year	UG	PG	Ph.D.	Total
2015-16	**-17**			**-17**
2016-17	-8			-8
2017-18	2			2
2018-19	16			16
2019-20	**34**			**34**
2020-21	36			36
2021-22	38			38
2022-23	39			39
2023-24	41			41
2024-25	**44**			**44**

Annx.4.7.8: Demand Projection of Food Technology Human Resources (Low Growth)

Year	UG	PG	Ph.D.	Total
2014-15	**38**	**14**	**14**	**66**
2015-16	39	14	14	68
2016-17	41	15	15	70
2017-18	42	15	15	72
2018-19	43	16	15	74
2019-20	**44**	**16**	**16**	**77**
2020-21	46	17	16	79
2021-22	47	17	17	81
2022-23	48	18	17	84
2023-24	50	18	18	86
2024-25	**51**	**19**	**18**	**89**

Annx.4.7.9 : Demand Projection of Food Technology Human Resources (Average Growth)

Year	UG	PG	Ph.D.	Total
2014-15	**38**	**14**	**14**	**66**
2015-16	40	15	14	69
2016-17	42	15	15	73
2017-18	44	16	16	76
2018-19	46	17	17	80
2019-20	**49**	**18**	**18**	**84**
2020-21	51	19	18	88
2021-22	54	20	19	93
2022-23	57	21	20	98
2023-24	59	22	21	102
2024-25	**62**	**23**	**22**	**108**

Annx.4.7.10: Demand Projection of Food Technology Human Resources (High Growth)

Year	UG	PG	Ph.D.	Total
2014-15	**38**	**14**	**14**	**66**
2015-16	41	15	15	71
2016-17	45	16	16	77
2017-18	48	18	17	83
2018-19	52	19	19	90
2019-20	**56**	**21**	**20**	**97**
2020-21	61	22	22	105
2021-22	66	24	24	113
2022-23	71	26	25	122
2023-24	76	28	27	132
2024-25	**83**	**30**	**30**	**142**

Annexure 4.8: Human Resources Stock Requirement Projections (demand) of Sericulture

Annx.4.8.1: Sericulture Human Resources Stock Requirement Projections for State sericulture Department (2%)

Year	Diploma	UG	PG	Ph.D.	Total
2014-15	15	240	111	7	358
2015-16	15	243	113	7	363
2016-17	15	247	114	7	369
2017-18	16	251	116	7	374
2018-19	16	255	118	8	380
2019-20	16	258	120	8	386
2020-21	16	262	121	8	391
2021-22	17	266	123	8	397
2022-23	17	270	125	8	403
2023-24	17	274	127	8	409
2024-25	17	278	129	8	415

Annx.4.8.2: Sericulture Human Resources Stock Requirement Projections for SAUs (5%)

Year	Diploma	UG	PG	Ph.D.	Total
2014-15	10	3	4	14	21
2015-16	11	3	4	14	22
2016-17	11	3	5	15	23
2017-18	12	4	5	16	24
2018-19	12	4	5	17	26
2019-20	13	4	5	17	27
2020-21	13	4	6	18	28
2021-22	14	4	6	19	30
2022-23	15	5	6	20	31
2023-24	16	5	7	21	33
2024-25	16	5	7	22	34

Annx.4.8.3: Sericulture Human Resources Stock Requirement Projections for Agro-chemical industries (2%)

Year	Diploma	UG	PG	Ph.D.	Total
2014-15	0	0	0	0	0
2015-16	0	0	0	0	0
2016-17	0	0	0	0	0
2017-18	0	0	0	0	0
2018-19	0	0	0	0	0
2019-20	0	0	0	0	0
2020-21	0	0	0	0	0
2021-22	0	0	0	0	0
2022-23	0	0	0	0	0
2023-24	0	0	0	0	0
2024-25	0	0	0	0	0

Annx. 4.8.4: Sericulture Human Resources Stock Requirement Projections or NGO (3%)

Year	Diploma	UG	PG	Ph.D.	Total
2014-15	0	0	0	0	0
2015-16	0	0	0	0	0
2016-17	0	0	0	0	0
2017-18	0	0	0	0	0
2018-19	0	0	0	0	0
2019-20	0	0	0	0	0
2020-21	0	0	0	0	0
2021-22	0	0	0	0	0
2022-23	0	0	0	0	0
2023-24	0	0	0	0	0
2024-25	0	0	0	0	0

Annx. 4.8.5: Sericulture Human Resources Stock Requirement Projections for Silkworm egg production industries (2%)

Year	Diploma	UG	PG	Ph.D.	Total
2014-15	15	4	1	0	5
2015-16	15	4	1	0	5
2016-17	16	4	2	0	5
2017-18	16	4	2	0	5
2018-19	16	4	2	0	5
2019-20	17	4	2	0	6
2020-21	17	4	2	0	6
2021-22	17	4	2	0	6
2022-23	18	4	2	0	6
2023-24	18	4	2	0	6
2024-25	18	4	2	0	6

Annx.4.8.6: Sericulture Human Resources Stock Requirement Projections for Financial Institutions

Year	Diploma	UG	PG	Ph.D.	Total
2014-15	0	0	0	0	0
2015-16	0	0	0	0	0
2016-17	0	0	0	0	0
2017-18	0	0	0	0	0
2018-19	0	0	0	0	0
2019-20	0	0	0	0	0
2020-21	0	0	0	0	0
2021-22	0	0	0	0	0
2022-23	0	0	0	0	0
2023-24	0	0	0	0	0
2024-25	0	0	0	0	0

Annx.4.8.7: Sericulture Human Resources Stock Requirement Projections for Self Employment (2%)

Year	Diploma	UG	PG	Ph.D.	Total
2014-15	2	1	1	0	2
2015-16	2	1	1	0	2
2016-17	2	1	1	0	2
2017-18	2	1	1	0	2
2018-19	2	1	1	0	2
2019-20	2	1	1	0	2
2020-21	2	1	1	0	2
2021-22	2	1	1	0	2
2022-23	2	1	1	0	2
2023-24	2	1	1	0	2
2024-25	2	1	1	0	2

Annx.4.8.8: Sericulture Human Resources Stock Requirement Projections for other sectors (2%)

Year	Diploma	UG	PG	Ph.D.	Total
2014-15	5	3	2	0	5
2015-16	5	3	2	0	5
2016-17	5	3	2	0	5
2017-18	5	3	2	0	5
2018-19	5	3	2	0	5
2019-20	6	3	2	0	6
2020-21	6	3	2	0	6
2021-22	6	3	2	0	6
2022-23	6	4	2	0	6
2023-24	6	4	2	0	6
2024-25	6	4	2	0	6

Annx.4.8.9: Total Demand Projected for Sericulture Human Resources Stock Requirement

Year	Diploma	UG	PG	Ph.D.	Total
2014-15	**47**	**251**	**119**	**21**	**391**
2015-16	48	255	121	22	398
2016-17	49	259	123	23	404
2017-18	51	263	125	23	411
2018-19	52	267	127	24	418
2019-20	**53**	**271**	**129**	**25**	**426**
2020-21	54	275	131	26	433
2021-22	56	280	134	27	441
2022-23	57	284	136	28	448
2023-24	59	288	138	30	456
2024-25	**60**	**293**	**140**	**31**	**464**

Annx.4.8.10: Projection of Demand Supply Gap of Sericulture Human Resources Stock Requirement

Year	Diploma	UG	PG	Ph.D.	Total
2015-16	**32**	**251**	**119**	**21**	**391**
2016-17	31	227	121	5	353
2017-18	31	231	123	5	359
2018-19	30	235	126	5	365
2019-20	**29**	**239**	**128**	**5**	**371**
2020-21	28	261	121	8	390
2021-22	27	274	127	8	409
2022-23	25	288	133	9	430
2023-24	23	302	140	9	451
2024-25	**21**	**318**	**147**	**9**	**474**

Annx.4.8.11: Projection of Demand Supply Gap of Sericulture Human Resources Stock Requirement (Low Growth)

Year	Diploma	UG	PG	Ph.D.	Total
2014-15	**47**	**251**	**119**	**21**	**391**
2015-16	48	258	123	22	403
2016-17	50	266	126	22	415
2017-18	51	274	130	23	427
2018-19	53	282	134	24	440
2019-20	**54**	**291**	**138**	**24**	**453**
2020-21	56	300	142	25	467
2021-22	58	309	147	26	481
2022-23	60	318	151	27	495
2023-24	61	327	155	27	510
2024-25	**63**	**337**	**160**	**28**	**525**

Annx.4.8.12: Projection of Demand Supply Gap of Sericulture Human Resources Stock Requirement (Average Growth)

Year	Diploma	UG	PG	Ph.D.	Total
2014-15	**47**	**251**	**119**	**21**	**391**
2015-16	49	263	125	22	411
2016-17	52	277	131	23	431
2017-18	54	290	138	24	453
2018-19	57	305	145	25	475
2019-20	**60**	**320**	**152**	**27**	**499**
2020-21	63	336	160	28	524
2021-22	66	353	168	29	550
2022-23	69	371	176	31	578
2023-24	73	389	185	32	607
2024-25	**77**	**409**	**194**	**34**	**637**

Annx. 4.8.13: Projection of Demand Supply Gap of Sericulture Human Resources Stock Requirement (High Growth)

Year	Diploma	UG	PG	Ph.D.	Total
2014-15	**47**	**251**	**119**	**21**	**391**
2015-16	51	271	129	23	422
2016-17	55	293	139	24	456
2017-18	59	316	150	26	493
2018-19	64	341	162	28	532
2019-20	**69**	**369**	**175**	**31**	**575**
2020-21	75	398	189	33	620
2021-22	81	430	204	36	670
2022-23	87	464	221	39	724
2023-24	94	502	238	42	782
2024-25	**101**	**542**	**257**	**45**	**844**

Annexure 4.9: Human Resources Stock Requirement Projections (demand) of Forestry

Annx. 4.9.1: Forestry Human Resources Stock Requirement Projections for State Forestry Department (1%)

Year	UG	PG	Ph.D.	Total
2014-15	850	393	25	1269
2015-16	859	397	26	1282
2016-17	867	401	26	1295
2017-18	876	405	26	1307
2018-19	885	409	26	1321
2019-20	894	413	27	1334
2020-21	903	418	27	1347
2021-22	912	422	27	1361
2022-23	921	426	27	1374
2023-24	930	430	28	1388
2024-25	939	435	28	1402

Annx. 4.9.2: Forestry Human Resources Stock Requirement Projections for SAUs (Forestry College) (5%)

Year	UG	PG	Ph.D.	Total
2014-15	4	5	17	26
2015-16	4	5	18	27
2016-17	4	6	19	29
2017-18	5	6	20	30
2018-19	5	6	21	32
2019-20	5	7	22	33
2020-21	5	7	23	35
2021-22	5	7	24	37
2022-23	6	8	25	38
2023-24	6	8	26	40
2024-25	6	8	28	42

Annx. 4.9.3: Forestry Human Resources Stock Requirement Projections Financial (2%)

Year	UG	PG	Ph.D.	Total
2014-15	0	1	0	1
2015-16	0	1	0	1
2016-17	0	1	0	1
2017-18	0	1	0	1
2018-19	0	1	0	1
2019-20	0	1	0	1
2020-21	1	1	0	1
2021-22	1	1	0	1
2022-23	1	1	0	1
2023-24	1	1	0	1
2024-25	1	1	0	1

Annx.4.9.4: Forestry Human Resources Stock Requirement Projections for Agro Processing (Timer & Wood Industry) (3%)

Year	UG	PG	Ph.D.	Total
2014-15	2	3	0	5
2015-16	2	3	0	5
2016-17	2	3	0	5
2017-18	2	3	0	5
2018-19	3	3	0	6
2019-20	3	3	0	6
2020-21	3	3	0	6
2021-22	3	3	0	6
2022-23	3	3	0	6
2023-24	3	4	0	7
2024-25	3	4	0	7

Annx. 4.9.5: Forestry Human Resources Stock Requirement Projections for NGO (2%)

Year	UG	PG	Ph.D.	Total
2014-15	1	1	0	2
2015-16	1	1	0	2
2016-17	1	1	0	2
2017-18	1	1	0	2
2018-19	1	1	0	2
2019-20	1	1	0	2
2020-21	1	1	0	2
2021-22	1	1	0	2
2022-23	1	1	0	2
2023-24	1	1	0	2
2024-25	1	1	0	2

Annx. 4.9.6: Forestry Human Resources Stock Requirement Projections for other sector (5%)

Year	UG	PG	Ph.D.	Total
2014-15	3	2	0	5
2015-16	3	2	0	5
2016-17	3	2	0	6
2017-18	3	2	0	6
2018-19	4	2	0	6
2019-20	4	2	0	6
2020-21	4	3	0	7
2021-22	4	3	0	7
2022-23	4	3	0	7
2023-24	5	3	0	8
2024-25	5	3	0	8

Annx. 4.9.7: Total Demand Projected for Forestry Human Resources Stock Requirement

Year	UG	PG	Ph.D.	Total
2014-15	**861**	**405**	**43**	**1308**
2015-16	870	409	44	1323
2016-17	879	414	45	1337
2017-18	888	418	46	1352
2018-19	897	423	47	1367
2019-20	**907**	**427**	**49**	**1383**
2020-21	916	432	50	1398
2021-22	926	437	51	1414
2022-23	935	442	53	1430
2023-24	945	447	54	1446
2024-25	**955**	**451**	**56**	**1463**

Annx. 4.9.8: Demand Supply Gap of Forestry Human Resources Stock Requirement Projections

Year	**UG**	**PG**	**Ph.D.**	**Total**
2015-16	720	378	39	1137
2016-17	729	383	40	1152
2017-18	738	387	42	1167
2018-19	747	392	43	1182
2019-20	756	397	44	1197
2020-21	842	390	25	1257
2021-22	884	409	26	1320
2022-23	928	430	28	1386
2023-24	975	451	29	1455
2024-25	1024	474	31	1528

Annx.4.9.9: Total Demand Projected for Forestry Human Resources Stock Requirement (Low Growth)

Year	UG	PG	Ph.D.	Total
2014-15	**861**	**405**	**43**	**1308**
2015-16	887	417	44	1348
2016-17	913	429	45	1388
2017-18	941	442	47	1430
2018-19	969	456	48	1473
2019-20	**998**	**469**	**49**	**1517**
2020-21	1028	483	51	1562
2021-22	1059	498	52	1609
2022-23	1090	513	54	1657
2023-24	1123	528	56	1707
2024-25	**1157**	**544**	**57**	**1758**

Annx.4.9.10: Total Demand Projected for Forestry Human Resources Stock Requirement (Average Growth)

Year	UG	PG	Ph.D.	Total
2014-15	**861**	**405**	**43**	**1308**
2015-16	904	425	45	1374
2016-17	949	446	47	1442
2017-18	997	469	49	1515
2018-19	1046	492	52	1590
2019-20	**1099**	**517**	**54**	**1670**
2020-21	1154	542	57	1753
2021-22	1211	570	60	1841
2022-23	1272	598	63	1933
2023-24	1335	628	66	2030
2024-25	**1402**	**659**	**69**	**2131**

Annx. 4.9.11: Total Demand Projected for Forestry Human Resources Stock Requirement (High Growth)

Year	UG	PG	Ph.D.	Total
2014-15	**861**	**405**	**43**	**1308**
2015-16	930	437	46	1413
2016-17	1004	472	50	1526
2017-18	1084	510	54	1648
2018-19	1171	551	58	1780
2019-20	**1265**	**595**	**63**	**1922**
2020-21	1366	642	68	2076
2021-22	1475	694	73	2242
2022-23	1593	749	79	2422
2023-24	1721	809	85	2615
2024-25	**1858**	**874**	**92**	**2825**

Annexure 4.10: Human Resources Stock Requirement Projections (demand) of Dairy Technology

Annex. 4.10.1: Dairy Technology Human Resources Stock Requirement Projections for KMF (5%)

Year	UG	PG	Ph.D.	Total
2014-15	**70**	**32**	**2**	**104**
2015-16	73	34	2	109
2016-17	77	36	2	115
2017-18	81	37	2	120
2018-19	85	39	3	126
2019-20	**89**	**41**	**3**	133
2020-21	93	43	3	139
2021-22	98	45	3	146
2022-23	103	48	3	154
2023-24	108	50	3	161
2024-25	**114**	**53**	**3**	169

Annex. 4.10.2: Dairy Technology Human Resources Stock Requirement Projections for SAUs (5%)

Year	UG	PG	Ph.D.	Total
2014-15	**3**	**4**	**14**	**22**
2015-16	3	5	15	23
2016-17	4	5	16	24
2017-18	4	5	17	25
2018-19	4	5	17	27
2019-20	4	6	18	28
2020-21	4	6	19	29
2021-22	5	6	20	31
2022-23	5	7	21	33
2023-24	5	7	22	34
2024-25	5	7	23	36

Annex. 4.10.3: Dairy Technology Human Resources Stock Requirement Projections for Animal feed industries (2%)

Year	UG	PG	Ph.D.	Total
2014-15	9	11	1	20
2015-16	9	11	1	20
2016-17	9	11	1	21
2017-18	10	12	1	21
2018-19	10	12	1	22
2019-20	10	12	1	22
2020-21	10	12	1	23
2021-22	10	13	1	23
2022-23	11	13	1	23
2023-24	11	13	1	24
2024-25	11	13	1	24

Annex. 4.10.4 : Dairy Technology Human Resources Stock Requirement Projections for Milk Processing & other industries (8%)

Year	UG	PG	Ph.D.	Total
2014-15	**18**	**7**	**0**	**25**
2015-16	19	8	0	27
2016-17	20	8	0	29
2017-18	22	9	0	31
2018-19	24	10	0	34
2019-20	26	11	0	37
2020-21	28	12	0	40
2021-22	30	12	0	43
2022-23	32	13	0	46
2023-24	35	14	0	50
2024-25	38	16	1	54

Annex. 4.10.5: Dairy Technology Human Resources Stock Requirement Projections for pharmaceuticals, drug industries and lab

Year	**UG**	**PG**	**Ph.D.**	**Total**
2014-15	4	2	0	6
2015-16	4	2	0	6
2016-17	5	2	0	7
2017-18	5	2	0	7
2018-19	6	2	0	8
2019-20	6	2	0	8
2020-21	6	3	0	9
2021-22	7	3	0	10
2022-23	7	3	0	10
2023-24	8	3	0	11
2024-25	8	3	0	12

Annex. 4.10.6: Dairy Technology Human Resources Stock Requirement Projections for NGOs (2%)

Year	**UG**	**PG**	**Ph.D.**	**Total**
2014-15	**4**	**1**	**0**	**5**
2015-16	4	1	0	5
2016-17	4	2	0	5
2017-18	4	2	0	5
2018-19	4	2	0	5
2019-20	4	2	0	6
2020-21	4	2	0	6
2021-22	4	2	0	6
2022-23	4	2	0	6
2023-24	4	2	0	6
2024-25	4	2	0	6

Annex.4.10.7: Dairy Technology Human Resources Stock Requirement Projections for Other sector (5%)

Year	UG	PG	Ph.D.	Total
2014-15	6	2	0	8
2015-16	6	2	0	8
2016-17	6	3	0	9
2017-18	6	3	0	9
2018-19	7	3	0	10
2019-20	7	3	0	10
2020-21	8	3	0	11
2021-22	8	3	0	11
2022-23	8	3	0	12
2023-24	9	4	0	12
2024-25	9	4	0	13

Annex.4.10.8: Total Demand of Dairy Technology Human Resources Stock Requirement Projections

Year	UG	PG	Ph.D.	Total
2014-15	**113**	**60**	**18**	**191**
2015-16	119	63	19	201
2016-17	125	66	20	211
2017-18	131	70	21	222
2018-19	138	73	22	233
2019-20	**146**	**77**	**23**	**245**
2020-21	153	80	24	258
2021-22	162	84	25	271
2022-23	170	89	26	285
2023-24	180	93	27	300
2024-25	**189**	**98**	**29**	**316**

Annex. 4.10.9: Projections of Demand Supply Gap of Dairy Technology Human Resources

Year	UG	PG	Ph.D.	Total
2015-16	**61**	**54**	**17**	**132**
2016-17	67	57	17	142
2017-18	74	60	18	153
2018-19	81	64	19	164
2019-20	**88**	**67**	**20**	**176**
2020-21	124	57	4	185
2021-22	130	60	4	194
2022-23	136	63	4	204
2023-24	143	66	4	214
2024-25	**150**	**70**	**4**	**224**

Annex. 4.10.10: Demand Projection of Dairy Technology Human Resource Requirements (low growth)

Year	UG	PG	Ph.D.	Total
2014-15	**113**	**60**	**18**	**191**
2015-16	116	62	18	197
2016-17	120	64	19	203
2017-18	123	66	19	209
2018-19	127	68	20	215
2019-20	***131***	***70***	***21***	***221***
2020-21	135	72	21	228
2021-22	139	74	22	235
2022-23	143	77	23	242
2023-24	147	79	23	249
2024-25	**152**	**81**	**24**	**257**

Annex. 4.10.11: Demand Projection of Dairy Technology Human Resources Requirement (Average Growth)

Year	UG	PG	Ph.D.	Total
2014-15	**113**	**60**	**18**	**191**
2015-16	118	63	19	201
2016-17	124	67	20	211
2017-18	131	70	21	221
2018-19	137	73	22	232
2019-20	**144**	**77**	**23**	**244**
2020-21	151	81	24	256
2021-22	159	85	25	269
2022-23	167	89	26	282
2023-24	175	94	28	296
2024-25	**184**	**98**	**29**	**311**

Annex.4.10.12: Demand Projection of Dairy Technology Human Resources Requirement (High Growth)

Year	UG	PG	Ph.D.	Total
2014-15	**113**	**60**	**18**	**191**
2015-16	122	65	19	206
2016-17	132	70	21	223
2017-18	142	76	22	241
2018-19	153	82	24	260
2019-20	**166**	**89**	**26**	**281**
2020-21	179	96	28	303
2021-22	193	104	31	327
2022-23	209	112	33	354
2023-24	225	121	36	382
2024-25	**243**	**130**	**38**	**412**

Annexure 4.11: Human Resources Stock Requirement Projections (demand) of Horticulture

Annx. 4.11.1: Horticulture Human Resources Stock Requirement Projections for Horticulture Departments/organization (5%)

Year	Diploma	UG	PG	Ph.D.	Total
2014-15	**35**	**474**	**219**	**14**	**707**
2015-16	37	497	230	15	742
2016-17	39	522	242	16	779
2017-18	41	548	254	16	818
2018-19	43	576	266	17	859
2019-20	**45**	**605**	**280**	**18**	**902**
2020-21	47	635	294	19	947
2021-22	49	667	308	20	995
2022-23	52	700	324	21	1045
2023-24	54	735	340	22	1097
2024-25	**57**	**772**	**357**	**23**	**1152**

Annx.4.11.2: Horticulture Human Resources Stock Requirement Projections for State Horticulture Universities (Teaching, Research and Extension) (5%)

Year	Diploma	UG	PG	Ph.D.	Total
2014-15	**40**	**46**	**62**	**200**	**308**
2015-16	42	49	65	210	323
2016-17	44	51	68	221	340
2017-18	46	53	71	232	357
2018-19	49	56	75	243	374
2019-20	**51**	**59**	**79**	**256**	**393**
2020-21	54	62	83	268	413
2021-22	56	65	87	282	433
2022-23	59	68	91	296	455
2023-24	62	72	96	311	478
2024-25	**65**	**75**	**100**	**326**	**502**

Annx. 4.11.3: Horticulture Human Resources Stock Requirement Projections for Financial Institutions (5.6 %)

Year	UG	PG	Ph.D.	Total
2014-15	**16**	**19**	**2**	**35**
2015-16	16	20	2	36
2016-17	17	21	2	38
2017-18	18	22	2	39
2018-19	18	23	2	41
2019-20	**19**	**23**	**2**	**43**
2020-21	20	24	2	44
2021-22	21	25	2	46
2022-23	22	26	2	48
2023-24	22	27	2	50
2024-25	**23**	**28**	**3**	**52**

Annx. 4.11.4 : Horticulture Human Resources Stock Requirement Projections for Seed and other private Sectors (5%)

Year	Diploma	UG	PG	Ph.D.	Total
2014-15	**60**	**30**	**41**	**4**	**75**
2015-16	63	32	43	4	79
2016-17	66	33	45	4	83
2017-18	69	35	48	4	87
2018-19	73	36	50	5	91
2019-20	77	38	53	5	96
2020-21	80	40	55	5	101
2021-22	84	42	58	5	106
2022-23	89	44	61	6	111
2023-24	93	47	64	6	116
2024-25	98	49	67	6	122

Annx. 4.11.5: Horticulture Human Resources Stock Requirement Projections for Horticulture Production sectors (5%)

Year	Diploma	UG	PG	Ph.D.	Total
2014-15	**40**	**19**	**14**	**2**	**35**
2015-16	42	20	15	2	37
2016-17	44	21	15	2	39
2017-18	46	22	16	2	41
2018-19	49	23	17	2	43
2019-20	**51**	**25**	**18**	**2**	**45**
2020-21	54	26	19	2	47
2021-22	56	27	20	2	49
2022-23	59	28	21	3	52
2023-24	62	30	22	3	54
2024-25	**65**	**31**	**23**	**3**	**57**

Annx. 4.11.6: Horticulture Human Resources Stock Requirement Projections for Processing of Fruits and Vegetables (10%)

Year	Diploma	UG	PG	Ph.D.	Total
2014-15	**30**	**18**	**22**	**2**	**40**
2015-16	32	20	24	2	44
2016-17	33	22	27	2	48
2017-18	35	24	29	3	53
2018-19	36	26	32	3	59
2019-20	**38**	**29**	**35**	**3**	**64**
2020-21	40	32	39	4	71
2021-22	42	35	43	4	78
2022-23	44	39	47	4	86
2023-24	47	42	52	5	94
2024-25	**49**	**47**	**57**	**5**	**104**

Annx. 4.11.6: Horticulture Human Resources Stock Requirement Projections for NGO's (5%)

Year	Diploma	UG	PG	Ph.D.	Total
2014-15	**20**	**5**	**5**	**0**	**10**
2015-16	21	5	5	0	11
2016-17	22	6	5	0	11
2017-18	23	6	6	0	12
2018-19	24	6	6	0	12
2019-20	**26**	**6**	**6**	**0**	**13**
2020-21	27	7	7	0	13
2021-22	28	7	7	0	14
2022-23	30	7	7	0	15
2023-24	31	8	8	0	16
2024-25	33	8	8	0	16

Annx. 4.11.7: Horticulture Human Resources Stock Requirement Projections for Self Employment (2%)

Year	Diploma	UG	PG	Ph.D.	Total
2014-15	**2**	**4**	**1**	**0**	**5**
2015-16	2	4	1	0	5
2016-17	2	4	2	0	5
2017-18	2	4	2	0	5
2018-19	2	4	2	0	5
2019-20	**3**	**4**	**2**	**0**	**6**
2020-21	3	4	2	0	6
2021-22	3	4	2	0	6
2022-23	3	4	2	0	6
2023-24	3	4	2	0	6
2024-25	**3**	**4**	**2**	**0**	**6**

Annx. 4.11.8: Horticulture Human Resources Stock Requirement Projections for Post-harvest Management and retailing (4%)

Year	Diploma	UG	PG	Ph.D.	Total
2014-15	**20**	**25**	**10**	**0**	**35**
2015-16	21	25	11	0	36
2016-17	22	26	11	0	38
2017-18	23	28	11	0	39
2018-19	24	29	12	0	41
2019-20	26	30	12	0	43
2020-21	27	31	13	0	44
2021-22	28	32	13	0	46
2022-23	30	34	14	0	48
2023-24	31	35	14	0	50
2024-25	33	36	15	1	52

Annx. 4.11.9: Horticulture Human Resources Stock Requirement Projections for other sectors (8%)

Year	Diploma	UG	PG	Ph.D.	Total
2014-15	**25**	**15**	**10**	**0**	**25**
2015-16	26	16	11	0	27
2016-17	28	17	11	0	29
2017-18	29	19	12	0	31
2018-19	30	20	13	0	34
2019-20	**32**	**22**	**14**	**0**	**37**
2020-21	34	24	15	0	40
2021-22	35	26	17	0	43
2022-23	37	28	18	0	46
2023-24	39	30	19	0	50
2024-25	**41**	**32**	**21**	**1**	**54**

Annx. 4.11.10: Horticulture Human Resources Stock Requirement Projections (Total)

Year	Diploma	UG	PG	Ph.D.	Total
2014-15	**270**	**647**	**402**	**224**	**1274**
2015-16	284	681	423	236	1340
2016-17	298	716	446	247	1409
2017-18	313	753	469	260	1482
2018-19	328	792	494	273	1559
2019-20	**345**	**833**	**521**	**287**	**1640**
2020-21	362	876	549	301	1726
2021-22	380	922	578	317	1816
2022-23	399	970	609	333	1911
2023-24	419	1020	642	349	2012
2024-25	**440**	**1074**	**677**	**367**	**2118**

Annx. 4.11.11: Demand Supply Gap in Horticulture Sector

Year	Diploma	UG	PG	Ph.D.	Total
2015-16	**227**	**398**	**291**	**193**	**881**
2016-17	233	394	299	197	888
2017-18	239	387	307	199	890
2018-19	245	377	315	199	887
2019-20	**250**	**361**	**322**	**198**	**876**
2020-21	254	340	329	194	858
2021-22	258	313	335	187	831
2022-23	260	278	341	177	792
2023-24	261	234	346	162	741
2024-25	**260**	**180**	**349**	**141**	**674**

Annx. 4.11.12: Horti 13 - Total Demand Projected (Low Growth)

Year	Diploma	UG	PG	Ph.D.	Total
2014-15	**270**	**647**	**402**	**224**	**1274**
2015-16	278	667	414	231	1312
2016-17	286	687	427	238	1351
2017-18	295	707	439	245	1392
2018-19	304	729	453	252	1434
2019-20	**313**	**751**	**466**	**260**	**1477**
2020-21	322	773	480	268	1521
2021-22	332	796	494	276	1567
2022-23	342	820	509	284	1614
2023-24	352	845	525	293	1662
2024-25	**363**	**870**	**540**	**301**	**1712**

Annx. 4.11.13: Horti 14 - Total Demand Projected (Average Growth)

Year	Diploma	UG	PG	Ph.D.	Total
2014-15	**270**	**647**	**402**	**224**	**1274**
2015-16	284	680	422	236	1337
2016-17	298	714	443	247	1404
2017-18	313	749	465	260	1475
2018-19	328	787	489	273	1548
2019-20	**345**	**826**	**513**	**286**	**1626**
2020-21	362	868	539	301	1707
2021-22	380	911	566	316	1792
2022-23	399	956	594	331	1882
2023-24	419	1004	624	348	1976
2024-25	**440**	**1055**	**655**	**365**	**2075**

Annx. 4.11.14: Horticulture Human Resources Stock Requirement Projections for Horticulture Sector (High Growth)

Year	Diploma	UG	PG	Ph.D.	Total
2014-15	**270**	**647**	**402**	**224**	**1274**
2015-16	292	699	434	242	1376
2016-17	315	755	469	262	1486
2017-18	340	816	506	283	1605
2018-19	367	881	547	305	1733
2019-20	**397**	**951**	**591**	**330**	**1872**
2020-21	428	1027	638	356	2021
2021-22	463	1110	689	384	2183
2022-23	500	1198	744	415	2358
2023-24	540	1294	804	448	2546
2024-25	**583**	**1398**	**868**	**484**	**2750**

Annexure 4.12: Human Resources Stock Requirement Projections (demand) of Veterinary and Animal Husbandry

Annx. 4.12.1: Veterinary and Animal Husbandry Human Resources Stock Requirement Projections for Development Department (2%)

Year	Diploma	UG	PG	Ph.D.	Total
2014-15	**25**	**588**	**272**	**18**	**878**
2015-16	26	600	278	18	896
2016-17	26	612	283	18	913
2017-18	27	624	289	19	932
2018-19	27	637	295	19	950
2019-20	28	**649**	**301**	**19**	969
2020-21	28	662	307	20	989
2021-22	29	676	313	20	1009
2022-23	29	689	319	21	1029
2023-24	30	703	325	21	1049
2024-25	30	**717**	**332**	**21**	1070

Annx. 4.12.2: Veterinary and Animal Husbandry Human Resources Stock Requirement Projections for SAUs (5%)

Year	Diploma	UG	PG	Ph.D.	Total
2014-15	**40**	**28**	**37**	**120**	**185**
2015-16	42	29	39	126	194
2016-17	44	31	41	133	204
2017-18	46	32	43	139	214
2018-19	49	34	45	146	225
2019-20	51	35	47	153	236
2020-21	54	37	50	161	248
2021-22	56	39	52	169	260
2022-23	59	41	55	178	273
2023-24	62	43	57	187	287
2024-25	65	45	60	196	301

Annx. 4.12.3: Veterinary and Animal Husbandry Human Resources Stock Requirement Projections for Poultry Sector (5%)

Year	Diploma	UG	PG	Ph.D.	Total
2014-15	**30**	**9**	**11**	**1**	**20**
2015-16	32	9	12	1	21
2016-17	33	10	12	1	22
2017-18	35	10	13	1	23
2018-19	36	11	13	1	24
2019-20	38	11	14	1	26
2020-21	40	12	15	1	27
2021-22	42	13	15	1	28
2022-23	44	13	16	1	30
2023-24	47	14	17	2	31
2024-25	49	15	18	2	33

Annx. 4.12.4: Veterinary and Animal Husbandry Human Resources Stock Requirement Projections for KMF (5%)

Year	Diploma	UG	PG	Ph.D.	Total
2014-15	**15**	**99**	**122**	**11**	**221**
2015-16	16	104	128	12	232
2016-17	17	110	134	12	244
2017-18	17	115	141	13	256
2018-19	18	121	148	13	269
2019-20	19	127	155	14	282
2020-21	20	133	163	15	296
2021-22	21	140	171	16	311
2022-23	22	147	180	16	327
2023-24	23	154	189	17	343
2024-25	24	162	198	18	360

Annx. 4.12.5: Veterinary and Animal Husbandry Human Resources Stock Requirement Projections for milk and meat processing industries (10%)

Year	Diploma	UG	PG	Ph.D.	Total
2014-15	**40**	**7**	**3**	**0**	**10**
2015-16	44	8	3	0	11
2016-17	48	8	4	0	12
2017-18	53	9	4	0	13
2018-19	59	10	4	0	15
2019-20	64	11	5	0	16
2020-21	71	12	5	0	18
2021-22	78	14	6	0	19
2022-23	86	15	6	0	21
2023-24	94	17	7	0	24
2024-25	104	18	8	0	26

Annx. 4.12.6: Veterinary and Animal Husbandry Human Resources Stock Requirement Projections for Veterinary Pharmaceuticals, Drug Industries and lab (5%)

Year	Diploma	UG	PG	Ph.D.	Total
2014-15	**15**	**12**	**8**	**0**	**20**
2015-16	16	13	8	0	22
2016-17	17	14	9	0	23
2017-18	19	15	10	0	25
2018-19	20	16	11	0	27
2019-20	22	18	11	0	29
2020-21	24	19	12	0	32
2021-22	26	21	13	0	34
2022-23	28	22	14	0	37
2023-24	30	24	16	0	40
2024-25	32	26	17	0	43

Annx. 4.12.7: Veterinary and Animal Husbandry Human Resources Stock Requirement Projections for animal feed industries (8%)

Year	Diploma	UG	PG	Ph.D.	Total
2014-15	**38**	**20**	**8**	**0**	**28**
2015-16	41	21	9	0	30
2016-17	44	23	9	0	33
2017-18	48	25	10	0	35
2018-19	52	27	11	0	38
2019-20	56	**29**	12	**0**	41
2020-21	60	30	13	1	44
2021-22	65	32	14	1	48
2022-23	70	35	15	1	52
2023-24	76	38	16	1	56
2024-25	82	**41**	18	**1**	60

Annx. 4.12.8: Veterinary and Animal Husbandry Human Resources Stock Requirement Projections for Private Veterinary service (3%)

Year	Diploma	UG	PG	Ph.D.	Total
2014-15	**35**	**60**	**39**	**1**	**100**
2015-16	37	63	41	1	105
2016-17	39	66	43	1	110
2017-18	41	69	45	1	116
2018-19	43	73	47	1	122
2019-20	45	**77**	**50**	**1**	128
2020-21	47	90	42	3	134
2021-22	49	94	44	3	141
2022-23	52	99	46	3	148
2023-24	54	104	48	3	155
2024-25	57	**109**	**50**	**3**	163

Annx. 4.12.9: Total Demand Projection for Veterinary and Animal Husbandry Human Resources Stock Requirement

Year	Diploma	UG	PG	Ph.D.	Total
2014-15	**239**	**828**	**506**	**152**	**1486**
2015-16	254	853	523	159	1536
2016-17	269	879	542	167	1588
2017-18	286	906	561	174	1642
2018-19	305	935	581	183	1698
2019-20	**324**	**964**	**602**	**191**	**1758**
2020-21	345	1003	614	202	1818
2021-22	367	1035	636	211	1883
2022-23	392	1069	660	221	1950
2023-24	417	1104	684	232	2020
2024-25	**420**	**971**	**502**	**224**	**1697**

Annx. 4.12.10: Projections of Demand Supply Gap of Veterinary and Animal Husbandry Human Resources Stock Requirement

Year	Diploma	UG	PG	Ph.D.	Total
2015-16	**151**	**590**	**437**	**115**	**1142**
2016-17	161	616	455	123	1194
2017-18	172	643	474	131	1248
2018-19	184	672	495	139	1305
2019-20	**196**	**701**	**516**	**147**	**1364**
2020-21	210	960	444	29	1432
2021-22	225	1008	466	30	1504
2022-23	241	1058	490	32	1579
2023-24	258	1111	514	33	1658
2024-25	**252**	**1166**	**540**	**35**	**1741**

Annx. 4.12.11: Projections of Demand of Veterinary and Animal Husbandry Human Resources Stock Requirement (Low Growth)

Year	Diploma	UG	PG	Ph.D.	Total
2014-15	**239**	**828**	**506**	**152**	**1486**
2015-16	246	853	521	157	1530
2016-17	253	878	536	161	1576
2017-18	261	905	553	166	1623
2018-19	269	932	569	171	1672
2019-20	**277**	**960**	**586**	**176**	**1722**
2020-21	285	989	604	181	1774
2021-22	294	1018	622	187	1827
2022-23	302	1049	641	193	1882
2023-24	312	1080	660	198	1938
2024-25	**321**	**1113**	**680**	**204**	**1997**

Annx. 4.12.12: Projections of Demand of Veterinary and Animal Husbandry Human Resources Stock Requirement (Average Growth)

Year	Diploma	UG	PG	Ph.D.	Total
2014-15	**239**	**828**	**506**	**152**	**1486**
2015-16	251	869	531	160	1560
2016-17	263	913	557	168	1638
2017-18	276	959	585	176	1720
2018-19	290	1006	615	185	1806
2019-20	**305**	**1057**	**645**	**194**	**1896**
2020-21	320	1110	678	204	1991
2021-22	336	1165	711	214	2090
2022-23	353	1223	747	225	2195
2023-24	370	1285	784	236	2305
2024-25	**389**	**1349**	**824**	**248**	**2420**

Annx. 4.12.13: Projections of Demand of Veterinary and Animal Husbandry Human Resources Stock Requirement (High Growth)

Year	Diploma	UG	PG	Ph.D.	Total
2014-15	**239**	**828**	**506**	**152**	**1486**
2015-16	258	894	546	164	1605
2016-17	278	966	590	177	1733
2017-18	301	1043	637	191	1871
2018-19	325	1127	688	207	2021
2019-20	**351**	**1217**	**743**	**223**	**2183**
2020-21	379	1314	802	241	2358
2021-22	409	1419	867	260	2546
2022-23	442	1533	936	281	2750
2023-24	477	1655	1011	304	2970
2024-25	**515**	**1788**	**1092**	**328**	**3207**

Annexure 4.13: Human Resources Stock Requirement Projections (Demand) for Fisheries

Annex. 4.13.1: Fishery Human Resources Stock Requirement Projections for Fisheries Department (5%)

Year	UG	PG	Ph.D.	Total
2014-15	**48**	**22**	**1**	**72**
2015-16	51	23	2	76
2016-17	53	25	2	79
2017-18	56	26	2	83
2018-19	59	27	2	88
2019-20	62	28	2	92
2020-21	65	30	2	96
2021-22	68	31	2	101
2022-23	71	33	2	106
2023-24	75	35	2	112
2024-25	79	36	2	117

Annex. 4.13.2: Fishery Human Resources Stock Requirement Projections for SAUs (Fisheries College) (5%)

Year	UG	PG	Ph.D.	Total
2014-15	**8**	**10**	**33**	**50**
2015-16	8	11	34	53
2016-17	8	11	36	55
2017-18	9	12	38	58
2018-19	9	12	40	61
2019-20	10	13	41	64
2020-21	10	13	44	67
2021-22	11	14	46	70
2022-23	11	15	48	74
2023-24	12	16	50	78
2024-25	12	16	53	81

Annex. 4.13.3: Fishery Human Resources Stock Requirement Projections for Boards and Corporation (2%)

Year	UG	PG	Ph.D.	Total
2014-15	**8**	**9**	**1**	**17**
2015-16	8	10	1	17
2016-17	8	10	1	18
2017-18	8	10	1	18
2018-19	8	10	1	18
2019-20	8	10	1	19
2020-21	9	11	1	19
2021-22	9	11	1	20
2022-23	9	11	1	20
2023-24	9	11	1	20
2024-25	9	11	1	21

Annex. 4.13.4: Fishery Human Resources Stock Requirement Projections for fish Processing plants (3%)

Year	UG	PG	Ph.D.	Total
2014-15	**6**	**8**	**1**	**15**
2015-16	6	8	1	15
2016-17	6	9	1	16
2017-18	7	9	1	16
2018-19	7	9	1	17
2019-20	7	10	1	17
2020-21	7	10	1	18
2021-22	7	10	1	18
2022-23	8	10	1	19
2023-24	8	11	1	20
2024-25	8	11	1	20

Annex. 4.13.5: Fishery Human Resources Stock Requirement Projections for Fish/Shrimp-Seed hatcheries and rearing units (5%)

Year	UG	PG	Ph.D.	Total
2014-15	**8**	**6**	**1**	**15**
2015-16	9	6	1	16
2016-17	9	7	1	17
2017-18	10	7	1	17
2018-19	10	7	1	18
2019-20	11	8	1	19
2020-21	11	8	1	20
2021-22	12	8	1	21
2022-23	12	9	1	22
2023-24	13	9	1	23
2024-25	13	10	1	24

Annex. 4.13.6: Fishery Human Resources Stock Requirement Projections for Fish feed Industries (3%)

Year	UG	PG	Ph.D.	Total
2014-15	**5**	**6**	**1**	**10**
2015-16	5	6	1	10
2016-17	5	6	1	11
2017-18	5	6	1	11
2018-19	5	6	1	11
2019-20	**5**	**6**	**1**	12
2020-21	8	4	0	12
2021-22	8	4	0	12
2022-23	8	4	0	13
2023-24	9	4	0	13
2024-25	**9**	**4**	**0**	13

Annex. 4.13.7: Fishery Human Resources Stock Requirement Projections for NGOs (2%)

Year	UG	PG	Ph.D.	Total
2014-15	**3**	**2**	**0**	**5**
2015-16	3	2	0	5
2016-17	3	3	0	5
2017-18	3	3	0	5
2018-19	3	3	0	5
2019-20	3	3	**0**	6
2020-21	3	3	0	6
2021-22	3	3	0	6
2022-23	3	3	0	6
2023-24	3	3	0	6
2024-25	3	3	**0**	6

Annex. 4.13.8: Fishery Human Resources Stock Requirement Projections for Financial Institutions

Year	UG	PG	Ph.D.	Total
2014-15	**4**	**1**	**0**	**5**
2015-16	4	2	0	6
2016-17	4	2	0	6
2017-18	5	2	0	7
2018-19	5	2	0	7
2019-20	**6**	**2**	**0**	8
2020-21	6	3	0	9
2021-22	7	3	0	10
2022-23	7	3	0	11
2023-24	8	4	0	12
2024-25	**9**	**4**	**0**	13

Annex. 4.13.9: Fishery Human Resources Stock Requirement Projections for Other Sectors (2%)

Year	UG	PG	Ph.D.	Total
2014-15	**3**	**2**	**0**	**5**
2015-16	3	2	0	5
2016-17	3	2	0	5
2017-18	3	2	0	5
2018-19	3	2	0	5
2019-20	**3**	**2**	**0**	6
2020-21	4	2	0	6
2021-22	4	2	0	6
2022-23	4	2	0	6
2023-24	4	2	0	6
2024-25	**4**	**2**	**0**	6

Annex. 4.13.10: Total Demand Fishery of Human Resources Stock Requirement Projections

Year	UG	PG	Ph.D.	Total
2014-15	**91**	**67**	**37**	**195**
2015-16	95	70	39	204
2016-17	100	73	41	213
2017-18	104	76	43	223
2018-19	109	79	45	233
2019-20	**114**	**82**	**47**	**243**
2020-21	122	83	49	254
2021-22	128	86	51	265
2022-23	134	90	54	277
2023-24	140	94	57	290
2024-25	**146**	**98**	**59**	**304**

Annex. 4.13.11: Projections of Demand Supply Gap in Fishery Human Resources Stock

Year	UG	PG	Ph.D.	Total
2015-16	**40**	**54**	**21**	**116**
2016-17	45	57	23	125
2017-18	49	60	25	134
2018-19	54	63	27	144
2019-20	**59**	**67**	**29**	**155**
2020-21	109	50	3	163
2021-22	114	53	3	171
2022-23	120	56	4	179
2023-24	126	58	4	188
2024-25	**132**	**61**	**4**	**198**

Annex. 4.13.12: Total Demand Projected for Fishery Human Resources Stock Requirement (Low Growth)

Year	UG	PG	Ph.D.	Total
2014-15	**91**	**67**	**37**	**195**
2015-16	94	69	38	201
2016-17	97	71	39	207
2017-18	100	74	40	213
2018-19	103	76	42	220
2019-20	**106**	**78**	**43**	**226**
2020-21	109	80	44	233
2021-22	112	83	45	240
2022-23	115	85	47	247
2023-24	119	88	48	255
2024-25	**122**	**90**	**50**	**263**

Annex. 4.13.13: Total Demand Projected for Fishery Human Resources Stock Requirement (Average Growth)

Year	UG	PG	Ph.D.	Total
2014-15	**91**	**67**	**37**	**195**
2015-16	96	71	39	205
2016-17	100	74	41	215
2017-18	106	78	43	226
2018-19	111	82	45	237
2019-20	**116**	**86**	**47**	**249**
2020-21	122	90	50	262
2021-22	128	95	52	275
2022-23	135	99	55	289
2023-24	141	104	57	303
2024-25	**148**	**110**	**60**	**318**

Annex.4.13.14: Total Demand Projected for Fishery Human Resources Stock Requirement (High Growth)

Year	UG	PG	Ph.D.	Total
2014-15	**91**	**67**	**37**	**195**
2015-16	98	73	40	211
2016-17	106	78	43	228
2017-18	115	85	47	246
2018-19	124	92	50	266
2019-20	**134**	**99**	**54**	**287**
2020-21	145	107	59	310
2021-22	156	115	63	335
2022-23	169	125	68	362
2023-24	182	134	74	391
2024-25	**197**	**145**	**80**	**422**